D0610217

Couper Institute Library
84 Clarkston Road
Glasgow G44 3DA
Phone: 0141 276 0771 Fax 276 0772

This book is due for return on or before the last date shown below. It may be renewed by telephone, personal application, fax or post, quoting this date, author, title and the book number

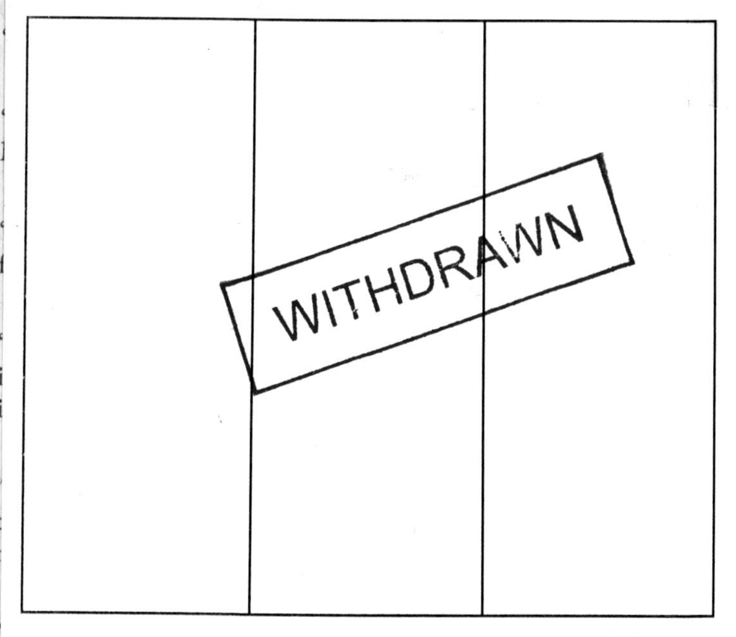

Glasgow Life and its service brands, including Glasgow Libraries, (found at www.glasgowlife.org.uk) are operating names for Culture and Sport Glasgow

engaging writing style.

—Jared Diamond, *The New York Times Book Review*

"McKibben is a mighty orator on the page here, just as he was in *The End of Nature* (1989) and *Eaarth* (2010), and his call for creating more compassionate and equitable societies is inspiring."

—***Pacific Standard***

"*Falter* is a bracing call to arms, one that concerned readers ignore at their peril."

—***Palo Alto Weekly***

"[An] unsettling look at the prospects for human survival. . . . Readers open to inconvenient and sobering truths will find much to digest in McKibben's eloquently unsparing treatise."

—***Publishers Weekly* (starred review)**

"[A] deeply caring, eloquently reasoned inquiry into environmental and techno-utopian threats . . . Profoundly compelling and enlightening, McKibben balances alarm with hope."

—***Booklist* (starred review)**

"A compelling call for change."

—***Kirkus Reviews***

"I braced myself to plunge into this book about the largest and grimmest of situations our species has faced, and then I found myself racing through it, excited by the grand synthesis of innumerable scientific reports on the details of the crisis. And then at the end I saw the book as a description of a big trap with a small exit we could take, if we take heed of what Bill McKibben tells us here and act on it."

—Rebecca Solnit, author of *A Paradise Built in Hell* and *Hope in the Dark*

"It's not an exaggeration to say that Bill McKibben has written a book so important, reading it might save your life, not to mention your home: Planet Earth. *Falter* is a brilliant, impassioned call to arms to save our climate from those profiting from its destruction before it's too late. Over and over, McKibben has proven one of the most farsighted and gifted voices of our times, and with *Falter* he has topped himself, producing a book that, honestly, everyone should read."

—Jane Mayer, bestselling author of *Dark Money*

FALTER

ALSO BY BILL McKIBBEN

Radio Free Vermont

Oil and Honey

The Global Warming Reader

Eaarth

American Earth

The Bill McKibben Reader

Fight Global Warming Now

Deep Economy

The Comforting Whirlwind

Wandering Home

Enough

Long Distance

Hundred Dollar Holiday

Maybe One

Hope, Human and Wild

The Age of Missing Information

The End of Nature

Falter

Has the Human Game Begun to Play Itself Out?

BILL McKIBBEN

Copyright © 2019 Bill McKibben

The right of Bill McKibben to be identified as the Author of
the Work has been asserted by him in accordance with the
Copyright, Designs and Patents Act 1988.

First published in 2019 by WILDFIRE
An imprint of HEADLINE PUBLISHING GROUP

First published in paperback in 2020 by WILDFIRE
An imprint of HEADLINE PUBLISHING GROUP

1

Apart from any use permitted under UK copyright law, this publication may
only be reproduced, stored, or transmitted, in any form, or by any means,
with prior permission in writing of the publishers or, in the case
of reprographic production, in accordance with the terms of
licences issued by the Copyright Licensing Agency.

Cataloguing in Publication Data is available from the British Library

ISBN 978 1 4722 6651 4

Offset in 10.34/14.44 pt Simoncini Garamond Std by Jouve (UK), Milton Keynes

Printed and bound in Great Britain by Clays Ltd, Elcograf S.p.A.

Headline's policy is to use papers that are natural, renewable and recyclable
products and made from wood grown in well-managed forests and other
controlled sources. The logging and manufacturing processes are expected
to conform to the environmental regulations of the country of origin.

HEADLINE PUBLISHING GROUP
An Hachette UK Company
Carmelite House
50 Victoria Embankment
London
EC4Y 0DZ

www.headline.co.uk
www.hachette.co.uk

For Koreti Tiumalu, 1975–2017—and for the thousands of other beloved colleagues who fight so hard for the planet's future

CONTENTS

FALTER

An Opening Note on Hope

Thirty years ago, in 1989, I wrote the first book for a wide audience on climate change—or, as we called it then, the greenhouse effect. As the title indicates, *The End of Nature* was not a cheerful book, and sadly its gloom has been vindicated. My basic point was that humans had so altered the planet that not an inch was beyond our reach, an idea that scientists underlined a decade later when they began referring to our era as the Anthropocene.

This volume is bleak as well—in some ways bleaker, because more time has passed and we are deeper in the hole. It offers an account of how the climate crisis has progressed and of the new technological developments in fields such as artificial intelligence that also seem to me to threaten a human future. Put simply, between ecological destruction and technological hubris, the human experiment is now in question. The stakes feel very high, and the odds very long, and the trends very ominous. So, I have no doubt that there are other books that would offer readers a merrier literary experience.

I know, too, that this bleakness cuts against the current literary grain. Recent years have seen the publication of a dozen high-profile books and a hundred TED talks devoted to the idea that everything in the world is steadily improving. They share not only a format (endless series of graphs showing centuries of decreasing infant mortality or rising income) but also a tone of perplexed exasperation that any thinking person could perceive the present moment as dark. As Steven Pinker, the author of the sanguine *Enlightenment Now*, explained, "None of us are as happy as we ought to be, given how amazing our world has become." People, he added, just "seem to bitch, moan, whine, carp and kvetch."[1]

I'm grateful for those books because, among other things, they remind us precisely how much we have to lose if our civilizations do indeed falter. But the fact that living conditions have improved in our world over the last few hundred years offers no proof that we face a benign future. That's because threats of a new order can arise—indeed, have now arisen. Just as a man or woman can grow in strength and size and wealth and intelligence for many years and then be struck down by some larger force (cancer, a bus), so, too, with civilizations. And—to kvetch and whine a little further—because of the way power and wealth are currently distributed on our planet, I think we're uniquely ill-prepared to cope with the emerging challenges. So far, we're *not* coping with them.

Still, there is one sense in which I am less grim than in my younger days. This book ends with the conviction that resistance to these dangers is at least possible. Some of that conviction stems from human ingenuity—watching the rapid spread of a technology as world-changing as the solar panel cheers me daily. And much of that conviction rests on events in my own life over the past few decades. I've immersed myself in movements working for change, and I helped found a group, 350.org, that grew into the first planetwide climate campaign. Though we haven't beaten the fossil fuel industry, we've organized demonstrations in every country on the globe save North Korea, and with our many colleagues around the world, we've won some battles. At the moment, we're helping as friends and colleagues push hard for a Green New Deal in the United States and similar steps around the world. (This book is dedicated to one of my dearest colleagues in that fight, Koreti Tiumalu, who died much too early, in 2017.) I've been to several jails, and to a thousand rallies, and along the way I've come to believe that we have the tools to stand up to entrenched power.

Whether that entrenched power can actually be beaten in time I do not know. A writer doesn't owe a reader hope—the only obligation is honesty—but I want those who pick up this volume to know that its author lives in a state of engagement, not despair. If I didn't, I wouldn't have bothered writing what follows.

PART ONE

The Size of the Board

1

If you viewed Earth from far above (and for better or worse, this book will often take a high, wide perspective), roofs would probably be the first feature of human civilization you'd notice. A descending alien would see many shapes, often corresponding to the local weather: A-frames for shedding snow, for instance. There are gambrel roofs, mansards, hipped and gabled roofs. Pagodas and other Asian temples often sport conical tops; Russian churches come with onion domes; Western churches sit beneath spires.

Palm leaves probably topped the earliest houses, but as humans began to grow grain in the Neolithic era, the leftover straw became a reliable roofing material. Some homes in Southern England have thatch roofs five hundred years old; new layers have been added over centuries till, in some cases, the roofs are seven feet thick. Though it is harder to find good stuff to work with—the introduction of short-stemmed wheat varieties and the widespread use of nitrogen fertilizer have weakened straw—thatch is now growing more popular with rich Europeans looking for green roofs; in Germany, for instance, you can now get a degree as a "journeyman specialist thatcher." But at least since the third century BC (perhaps beginning with Greek temples deemed valuable enough to protect from fire) humans have been tending toward hard roofs. Terra-cotta tiles spread rapidly around the Mediterranean and to Asia Minor; slate roofs became popular for their low maintenance; where trees are plentiful, wood shakes and slabs of bark work well. Given that the average human being currently resides in an urban slum, it is possible that corrugated iron shelters more sleepers than anything else.

Do you find this a little dull? Good. What I want to talk about is the human game—the sum total of culture and commerce and politics; of religion and sport and social life; of dance and music; of dinner and art and cancer and sex and Instagram; of love and loss; of everything that comprises the experience of our species. But that's beyond my powers, at least till I'm warmed up. So, I've looked for the most mundane aspect of our civilization I can imagine. Almost no one thinks about her roof from one year's end to another, not unless it springs a leak. It's a given. And so, it will illustrate my point—even the common and boring roof demonstrates the complexity, the stability, and the reach of this human game.

Consider the asphalt shingle, which tops most homes in the West and is itself, doubtless, the dullest of all forms of roofing. The earliest examples date to 1901, and the first manufacturer was the H.M. Reynolds Company of Grand Rapids, Michigan, which sold its product under the slogan "The Roof That Stays Is the Roof That Pays." Asphalt occurs naturally in a few places on Earth—the tar sands of Alberta, for instance, are mostly bitumen, which is the geologist's word for asphalt. But the asphalt used in shingles comes from the oil-refining process: it's the stuff that still hasn't boiled at five hundred degrees Fahrenheit. Vacuum distillation separates it from more valuable products such as gasoline, diesel, and naphtha; it then is stored and transported at high temperatures until it can be used, mostly for making roads. But some of it is diverted to the plants that make shingles, where manufacturers add granules of some mineral (slate, fly ash, mica) to improve durability. The CertainTeed Corporation, the world's biggest shingle manufacturer, has produced a video showing what it rightly calls "this underappreciated process" at its plant in Oxford, North Carolina, one of sixty-one facilities it operates around the country. The video shows a ballet of pouring and dumping and conveying, as limestone arrives by rail car to be crushed and mixed with hot asphalt and then coated onto hundreds of thousands of miles of fiberglass mat. A thin mist of water is sprayed, and as it evaporates, the sheet cools, ready to be cut and then bundled onto pallets in a giant warehouse, to await distribution.[1]

Marvel for a moment at the thousands of events that must synchronize for all this to work: the oil drilled (maybe deep undersea, or in the equatorial desert); the pipelines and rail lines laid; the refineries constructed (and at each step, the money raised). The limestone and the sand need mining, too, and the miles of fiberglass net must be fabricated on some other production line. The raw materials are all sucked into the North Carolina factory, and then the finished shingles must be spewed back out again, across rail lines and truck routes and into a network of building supply stores, where contractors can haul them to building sites, confident that they've been rated for resistance to wind, fire, and discoloration. Think, again, of the sheer amount of human organization required for the American Society for Testing and Materials to produce directive D3462-87 ("Asphalt Shingles Made from Glass Felt and Surfaced with Mineral Granules") and then to enforce its mandates.

We could, clearly, repeat this exercise for everything you see around you, and everything you hear, and everything you smell—all the infinitely more interesting activities always under way beneath all those roofs. As I write, for instance, I'm listening to Orchestra Baobab on Spotify. It was the house band at a Dakar nightclub in the 1970s, where its music reflected the Cuban beats that came with sailors to West Africa in the 1940s; eventually the group recorded its best album at a Paris studio, and now it somehow resides on a computer server where 196,847 people from across the planet listen to it each month. Try to parse the play of history and technology and commerce and spirituality and swing that make up the sound pouring into my headphones—the colonialisms layered on top of one another; the questions of race, identity, pop, purity. Or consider what I'm going to have for dinner, or what you're wearing on your back—*everything* comes with strings attached, and you can follow those strings into every corner of our past and present.

What I'm calling the human game is unimaginably deep, complex, and beautiful. It is also endangered. Indeed, it is beginning to falter even now.

I'll spend this book explaining that danger and, at the end, pointing to some ways we might yet avert it. But I think it's best to begin by stressing not the shakiness of the human game but, instead, its stability. For humans, all of us together, have built something remarkable, something we rarely stand back and simply acknowledge. The sum of the projects of our individual lives, the total of the institutions and enterprises we have created, the aggregate of our wishes and dreams and labors, the entirety of our ceaseless activity—it is a wonder. I call it a game because it has no obvious end. Like any game, it doesn't really *matter* how it comes out, at least in the largest sense of Our Place in the Universe, and yet, like any game, it absorbs the whole concentration of those involved. And even if it has no ultimate aim that doesn't mean it lacks rules, or at least an aesthetic: by my definition, the game is going well when it creates more dignity for its players, and badly when that dignity diminishes.

Dignity, in the context of the human game, can be measured in many ways: enough calories, freedom from fear, clothes to wear, useful work. And by plenty of those measures, we're on a roll. Extreme poverty (life on two dollars a day or less) is far rarer than it used to be. Many of the diseases that poverty helped spread have lessened, too: worms in your gut, say. Even compared to the twentieth century, violence is now far less likely to kill us—of the more than 55 million people who died around the world in 2012, war killed just 120,000 of them.[2] Eighty-five percent of adults can read now, a staggering increase inside two generations.[3] Women, with more education and at least a modicum of equality, have gone from having more than five kids apiece on average in 1970 to having fewer than two and a half today, probably the most rapid and remarkable demographic change the planet has ever witnessed. In the year 1500, humans managed to produce goods and services worth $250 billion in today's dollars—five hundred years later, that number is $60 trillion, a 240-fold increase.[4] The chorus of affirmation swells, from Steven Pinker insisting we're in an age of unprecedented enlightenment to Donald Trump tweeting, "There is an incredible spirit of optimism sweeping the country right now—we're bringing back the JOBS!"

We're quite accustomed to this idea of progress, so accustomed that some can't imagine anything else: the former chief economist of the World Bank, Kaushik Basu, recently predicted that, in fifty years, global GDP will be growing 20 percent a year, meaning that income and consumption will be doubling every four years or so.[5] There are, each day, more ideas hatched, more songs sung, more pictures taken, more goals scored, more schoolbooks read, more money invested.

And yet. There are other authorities almost as highly placed as the former chief economist of the World Bank. Pope Francis, in his landmark 2015 encyclical on the environment and poverty, said, "The earth, our home, is beginning to look more and more like an immense pile of filth." Don't consider popes sufficiently authoritative? Consider this: In November 2017, fifteen thousand scientists from 184 countries issued a stark "warning to humanity." Just like Pinker, they had charts, but theirs depicted everything from the decline in freshwater per person to the spread of anaerobic "dead zones" in the world's seas. As a result, the scientists predicted, we face "widespread misery and catastrophic biodiversity loss"; soon, they added, "it will be too late to shift course away from our failing trajectory." (Within six months, that warning was already the sixth-most-discussed academic paper in history.)[6] The worries have grown severe enough that a NASA-funded group recently created the Human and Nature DYnamics (HANDY) program to model the fall of the Roman, Han, Mauryan, and Gupta Empires, and when they pushed the button, it spit out a disquieting forecast: "Global industrial civilization could collapse in coming decades due to unsustainable resource exploitation and increasingly unequal wealth distribution." (The fact that I'd never even heard of the Mauryan Empire gave me a quiet shiver.) In this model, by the way, one of the greatest dangers came from elites who argued against structural change on the grounds that "so far" things were working out.[7]

That "so far" is always the problem, as the man who fell off the skyscraper found out. If you want to fret, you can find plenty of indications that the pavement is approaching with discouraging speed. A third of the planet's land is now severely degraded, with "persistent declining trends in productivity," according to a September 2017 report.[8] We've displaced most everything else: if you weigh the earth's terrestrial vertebrates, humans account for 30 percent of their total mass, and our farm animals for another 67 percent, meaning wild animals (all the moose and cheetahs and wombats combined) total just 3 percent.[9] In fact, *there are half as many wild animals on the planet as there were in 1970*, an awesome and mostly unnoticed silencing. In 2018, scientists reported that the planet's oldest and largest trees were dying fast, "as climate change attracts new pests and diseases to forests." The baobab—Africa's tree of life, in whose shade people first hunted and gathered—can live as long as 2,500 years, but five of the six oldest specimens on the planet have died in the last decade.[10] Before century's end, climate change may kill off the cedars of Lebanon—plundered by Gilgamesh, name-checked in the Bible—as snow cover disappears and sawflies hatch earlier in the heat.[11]

Even our arks are leaking: with a burst of foresight, the world's agronomists designed a Global Seed Vault in an Arctic mountain, an impregnable bank where they could save a million varieties of seed covering all the Earth's important food crops. Eight years after it opened, during the hottest year ever recorded on the planet, melting snow and heavy rain flooded the entrance tunnel and then froze. The seeds weren't damaged, but the builders were no longer confident that they'd constructed a stronghold that would last into deep time. "It was not in our plans to think the permafrost would not be there and that it would experience extreme weather like that," a Norwegian government spokesman said.[12]

And yet nothing slows us down—just the opposite. By most accounts, we've used more energy and resources during the last thirty-five years than in all of human history that came before.[13] Every economic assumption our governments make about the future requires doubling the size of the economy again, and then again, and then again during the lives of the youngest people on the planet. So, it's hard to make the argument that past performance indicates much about the future—it looks like the same game, but it's on new ground.

In part, that's because the past is so short. We are the first acutely self-conscious species, so wrapped up in our own story that we rarely stop to remember how short that story really is. Day to day, we forget that if the billions of years of life on Earth were scaled to a twenty-four-hour day, our settled civilizations began about a fifth of a second ago.[14] That short burst covers the taming of fire, the development of language, the rise of agriculture. On the time scale of a human life, these changes seemed to take forever, but in geological reality, they occupied the blink of an eye. And now we see shifts (the development of nuclear weapons, the rise of the internet) that change many of our assumptions in real time. So, the fact that even over this short span we've seen the routine and often sudden collapse of one civilization after another might give us pause. And in some ways, it does—books such as Jared Diamond's *Collapse* intrigue us with their stories of past calamities, from Greenland to Easter Island.

But these warnings also somehow seem to give us confidence, because, after all, things continued. Rome fell, and something else rose. The Fertile Crescent turned to desert, but we found other places to grow our food. The cautionary tales about transcending our limits (the apple in Eden, the Tower of Babel, Icarus) seem silly to us because we're still here, and we keep transcending one limit after another.

Sometimes we scare ourselves for a season, but then we shake it off. As the postwar explosion in consumption spread across much of the planet, for instance, modern environmentalism also took shape, questioning whether this trajectory was sustainable. That movement reached its first height in 1972, with the publication of a slim book called *The Limits to Growth.* Without specifying precisely how and when, the authors of that book, and the computer models they built, predicted that our pell-mell growth would, "sometime within the next hundred years," collide with many natural limits, and that without dramatic change, "the most probable result will be a rather sudden and uncontrollable decline in both population and industrial capacity." Alternately, they said, the nations of the world could "create a condition of ecological and economic stability that is sustainable far into the future," a task that would be easier the sooner we began.[15] Needless to say, we've not done that. Though we've taken the environmental idea semi-seriously, passing the laws that cleaned air and water, we've never taken it anywhere near as seriously as we've taken further growth. On his way to the theoretically groundbreaking Rio environmental summit in 1992, the first President Bush famously declared, "The American way of life is not up for negotiation,"[16] and as it turns out, he was correct—and speaking for much of the world. And so far, we've gotten away with it: even as we keep accelerating, the game spins on.

So, why should you take seriously my fear that the game, in fact, may be starting to play itself out? The source of my disquiet can be summed up in a single word, a word that will be repeated regularly in this book: *leverage.* We're simply so big, and moving so fast, that every decision carries enormous risk.

Rome's collapse was, of course, a large-ish deal. But given that there were vast swaths of the world that didn't even know there *was* a Roman Empire, it wasn't a big deal *everywhere.* Rome fell, and the Mayans didn't tremble, nor the Chinese, nor the Inuit. But an interconnected world is different. It offers a certain kind of stability—everyone in every country can all hear the scientists warning of impending climate change, say—but it removes the defense of distance. And the sheer size of our consumption means we have enormous leverage of a different sort—no Roman emperor could change the pH of the oceans, but we've managed that trick in short order. And, finally, the new scale of our technological reach amplifies our power in extraordinary ways: much of this book will be devoted to examining the godlike powers that come with our rapid increases in computing speed, everything from human genetic engineering to artificial intelligence.

We are putting the human game at risk, that is, from things going powerfully wrong and powerfully right. As we shall see, humans have now emerged as a destructive geologic force—the rapid degradation of the planet's physical systems that was still theoretical when I wrote *The End of Nature* is now under way. Indeed, it's much farther advanced than most people realize. In 2015, at the Paris climate talks, the world's governments set a goal of holding temperature increases to 1.5 degrees Celsius and, at the very least, below 2 degrees; by the fall of 2018 the IPCC reported that we might go past that 1.5 degree mark by 2030. That is to say, we will have drawn a line in the sand and then watched a rising tide erase it, all in a decade and a half.

And humans have simultaneously emerged as a massive *creative* force, in ways that threaten the human game not through destruction but through substitution. Robots are not just another technology, and artificial intelligence not just one more improvement like asphalt shingles. They are instead a replacement technology, and the thing's that's going obsolete may well be us. If we're not humans, then the human game makes no sense.

Over our short career as a species, human history has risen and fallen, gotten stuck and raced ahead, stagnated and flourished. Only now, though, have we achieved enough leverage that we can bring it to an end, both by carelessness and by design. As a team of scientists pointed out recently in *Nature*, the physical changes we're currently making by warming the climate will "extend longer than the entire history of human civilization thus far."[17] And as the Israeli historian and futurist Yuval Harari recently wrote, "Once technology enables us to re-engineer human minds, *Homo sapiens* will disappear, human history will come to an end, and a completely new process will begin, which people like you and me cannot comprehend."[18] That is to say, the game that we've been playing may end with neither a bang nor a whimper, but with the burble of a rising ocean and the soft beep of some digital future being birthed.

The outsize leverage is so crucial because, for the first time, we threaten to cut off our own lines of retreat. When Rome fell, something else was there. We had, to draw on pinball, perhaps the most delightfully pointless of games, another silver ball, another chance. But our current changes are so big that they're starting to tilt the whole machine, at which point it will fall silent. And as we shall see, because of the radical inequality we've allowed to overtake our society, the key decisions have been and will be made by a handful of humans in a handful of places: oil company executives in Houston, say, and tech moguls in Silicon Valley and Shanghai. Particular people in particular places at a particular moment in time following a particular philosophic bent: that's leverage piled on top of leverage. And their ability to skew our politics with their wealth is one more layer of leverage. It scares me.

It scares me even though the human game is not perfect—in fact, no one gets out of it alive, and no one without sadness and loss. For too many people, it's much more tragic than it needs to be—indeed, it's wretched, and often because its rules have been rigged to favor some and damage others. Given that I've been in the luckier fraction, the game may seem more appealing to me than to others. And perhaps its loss will not feel as acute to those being born now: certainly, they will not mourn the absence of things they did not know, just as we are not wrenched by the loss of the dinosaurs. If you back up far enough, it's possible to be philosophical about anything—the sun is going to blow up eventually, after all. But that's more philosophy than I can manage;

for me, and for many others, the loss of this game is the largest conceivable tragedy, if, indeed, we can conceive it.

And so, we will fight—some of us already are fighting. And we can, I think, see some of the ways out, even if the odds of their succeeding are not great. Success would require real changes in thinking from both conservatives and progressives. (Conservatives, oddly, tend not to worry about conservation; progressives tend to think all progress is good.) But if those changes came fast enough, the game could roll on: scientists estimate that we have five billion years until the sun turns into a red giant and expands past Earth's orbit. I'm neither optimistic nor pessimistic, just realistic—enough to know engagement is our only chance.

I said before that the human game we've been playing has no rules and no end, but it does come with two logical imperatives. The first is to keep it going, and the second is to keep it human.

2

To walk the roads through even a corner of Alberta's vast tar sands complex is to visit a kind of hell. This may be the largest industrial complex on our planet—the largest dam on Earth holds back one of the many vast settling "ponds," where sludge from the mines combines with water and toxic chemicals in a black soup. Because any bird that landed on the filthy water would die, cannons fire day and night to scare them away. If you listen to the crack of the guns, and to the stories of the area's original inhabitants, whose forest was ripped up for the mines, you understand that you are in a war zone. The army is mustered by the Kochs (the biggest leaseholders in the tar sands) and ConocoPhillips and PetroChina and the rest, and their enemy is all that is wild and holy. And they are winning.

It is hideous, a vandalism of the natural and human world that can scarcely be imagined. I've spent years working to end it, and my efforts have been small compared to the unending fight of the people who live there. And yet, giant as this scar is, in itself it represents no real threat to the human game. The Earth is not infinite, but it is very large, and if you retreat far enough, even this scab (the single ugliest sight I've witnessed in a lifetime of traveling the planet) gets swallowed up in the vastness that is Canada's boreal forest, and that in the vastness of North America, and that in the vastness of the hemisphere.

Likewise, to wake up in Delhi at the moment is to wake up in a gray, grim purgatory. The clatter and smell of one of the planet's most crowded cities assail you as always, but some days the smog grows so thick you can't see the end of the block. Walking down the pavement, you seem almost alone, and the city noise seems as if it must be made by ghosts. When the air is at its worst, when the smoke from the region's farms burning off stubble combines with the exhaust of cars and buses and the cooking fires of the slums, it's almost unbearable: in one recent outbreak, the international airlines scrubbed their flights into Delhi because the runway was invisible, and then cars began crashing on the highways, and then the city's trains were cancelled due to poor visibility. Imagine how bad the air must be to cancel a train, which runs on a track. At a big international cricket match the next month, with pollution levels fifteen times the global standard for safety, players began "continuously vomiting." After halting play for twenty minutes, the umpire said, "There aren't too many rules regarding pollution."[1]

Delhi's air pollution may currently be the worst in the world, besting even the smoke-racked Chinese cities where the authorities installed giant LED screens to show video of the sun rising. Or maybe Lahore, in Pakistan, deserves the crown: particulate levels there have reached thirty times the safe level, producing a soupy brown haze likened by one journalist to a "giant airport smokers' lounge."[2] Asian authorities routinely close schools because of the bad air, but because most homes lack filters, that doesn't help much. A large-scale study found that, of the 4.4 million children in Delhi, fully half had irreversible lung damage from breathing the air.[3] Around the world, pollution kills 9 million people a year, far more than AIDS, malaria, TB, and warfare combined.[4] In the worst years, a third of the deaths in China can be blamed on smog, and by 2030, it may claim 100 million victims worldwide.[5]

It is sick, sad, unnecessary—the biggest public health crisis on the planet. And yet, even it represents no *existential* threat to the human game. If the devastation of the tar sands is limited in space, this assault is limited in time. It can and will be solved, too slowly, with far too much human anguish, but that is the lesson from London, from Los Angeles, even from Beijing, which has begun, haltingly, to clear its air.

The list of such severe environmental problems grows ever longer: dead zones in the oceans where fertilizer pours off farms along with irreplaceable topsoil; great gyres of plastic waste spinning in the seas; suburbs spilling across agricultural land, and agricultural land overrunning tropical forest; water tables quickly sinking as aquifers drain. These issues rightly demand, and even rightly monopolize, our attention because the threats they represent are so stark and so immediate. And yet, one imagines that we will survive them as a species, impoverished in many ways, but not threatened in our basic existence. People, and other creatures, will be robbed of dignity—they're all signs of a game badly played—but the game goes on.

But not every threat is like that. There's a small category (a list with three items) of physical threats so different in quantity that they become different in quality, their effects so far-reaching that we can't be confident of surviving them with our civilizations more or less intact. One is large-scale nuclear war; it's always worth recalling J. Robert Oppenheimer's words as he watched the first bomb test, quoting from Hindu scripture: "Now I am become Death, the destroyer of worlds." So far, the cobbled-together and jury-rigged international efforts to forestall an atomic war have worked, and indeed, for much of the last fifty years those safeguards, formal and informal, have seemed to be strengthening. That we have nuclear nightmares again is mostly testimony to the childishness of President Trump and his pal in North Korea—they seem nearly alone in not understanding "why we can't use them."

Second on that list of threats is the small group of chemicals that, just in time, scientists discovered were eroding the ozone layer, a protective shield that 99 percent of us didn't even know existed. Had those scientists not sounded the alarm, we would have walked blindly off a cliff—literally, in many cases, as cataracts are one of the most common symptoms of being bathed in the ultraviolet radiation that the ozone layer blocks. Within a decade, the chemical companies had ceased their obstruction and the Montreal Protocol began removing chlorofluorocarbons from the atmosphere. The ozone hole over the Antarctic now grows smaller with each decade, and now scientists expect it will be wholly healed by 2060.

And the third, of course, is climate change, perhaps the greatest of all these challenges, and certainly the one about which we've done the least. It may not be quite game-ending, but it seems set, at the very least, to utterly change the board on which the game is played, and in more profound ways than almost anyone now imagines. The habitable planet has literally begun to shrink, a novel development that will be the great story of our century.

Climate change has become such a familiar term that we tend to read past it—it's part of our mental furniture, like *urban sprawl* or *gun violence*. So, let's remember exactly what we've been up to, because it should fill us with awe; it's by far the biggest thing humans have ever done. Those of us in the fossil fuel–consuming classes have, over the last two hundred years, dug up immense quantities of coal and gas and oil, and burned them: in car motors, basement furnaces, power plants, steel mills. When we burn them, the carbon atoms combine with oxygen atoms in the air to produce carbon dioxide. The molecular structure of carbon dioxide traps heat that would otherwise have radiated back out to space. We have, in other words, changed the energy balance of our planet, the amount of the sun's heat that is returned to space. Those of us who burn lots of fossil fuel have changed the way the world operates, fundamentally.

The scale of this change is the problem. If we just burned a little bit of fossil fuel, it wouldn't matter. But we've burned enough to raise the concentration of carbon dioxide in the atmosphere from 275 parts per million to 400 parts per million in the course of two hundred years. We're on our way, on the present trajectory, to 700 parts per million or more. Because none of us knows what a "part per million" feels like, let me put it in other terms. The extra heat that we trap near the planet because of the carbon dioxide we've spewed is equivalent to the heat from 400,000 Hiroshima-sized bombs every day, or four each second.[6] As we will see, this extraordinary amount of heat is wreaking enormous changes, but for now, don't worry about the effects; just marvel at the magnitude: the extra carbon released to date, if it could be amassed in one place, would form a solid graphite column twenty-five meters in diameter that would stretch from here to the moon.[7] There are perhaps four other episodes in Earth's 4.5-billion-year history where carbon dioxide has poured into the atmosphere in greater volumes, but never at greater speeds—right now we push about forty billion tons into the atmosphere annually. Even during the dramatic moments at the end of the Permian Age, when most life went extinct, the carbon dioxide content of the atmosphere grew at perhaps one-tenth the current pace.[8]

The results, already, have been extraordinary. In the thirty years I've been working on this crisis, we've seen all twenty of the hottest years ever recorded. So far, we have warmed the earth by roughly two degrees Fahrenheit, which in a masterpiece of understatement the *New York Times* once described as "a large number for the surface of an entire planet."[9] This is humanity's largest accomplishment, and indeed the largest thing any one species has ever done on our planet, at least since the days two billion years ago when cyanobacteria (blue-green algae) flooded the atmosphere with oxygen, killing off much of the rest of the archaic life on the planet. "Faster than expected" is the watchword of climate scientists—the damage to ice caps and oceans that scientists (conservative by nature) predicted for the end of the century showed up decades early. "I've never been at a climate conference where people say 'that happened slower than I thought it would,'" one polar expert observed in the spring of 2018.[10] At about the same time, a team of economists reported that there was a 35 percent chance that the United Nations' previous "worst-case scenario" for global warming was in fact too optimistic.[11] In January 2019 scientists concluded the Earth's oceans were warming 40 percent faster than previously believed.

"We are now truly in uncharted territory," said the director of the World Meteorological Organization in the spring of 2017, after final data showed that the previous year had broken every heat record.[12] He was speaking literally, not metaphorically—*we were off the actual charts.* That summer, an Atlantic hurricane developed well to the east of where any such storm had ever been seen before. Instead of crashing into Mexico and Louisiana and Florida, it spent its fury on Ireland and Scotland. When the National Oceanic and Atmospheric Administration showed the storm forecast on its computerized maps, the image looked odd: the cone of winds stopped abruptly in a straight line at latitude 60 degrees north—because, it turns out, it had never occurred to the people programming the forecasting models that a hurricane would reach that line. "That's a pretty unusual place to have a tropical cyclone," the programmer said. "Maybe that's something we'll have to go back and revisit what the boundary is."[13] Maybe so.

If you find a stout enough man, you can give him a pretty hefty shove and not much happens (unless, with some justification, he gets mad). When the global warming era began, we did not know how stout the planet was—it was possible that its systems would tolerate a lot of pushing without much change. The earth seems, after all, like a robust place: its ice sheets are miles thick, its oceans miles deep. But the lesson of the last thirty years is unequivocal: the planet was actually finely balanced, and the shove we've given it has knocked it very much askew. Let's look for a long minute at what has happened so far, remembering always that we're still in the early stages of global warming and that things will proceed inevitably from worse to worse yet and then keep on going.

Consider something fairly simple: the planet's hydrology, the way water moves around the earth. Water evaporates off the surface of the earth and the ocean, and then falls as rain and snow, an endless pump for keeping the earth's essential fluid in constant motion. But if you increase the amount of heat (of energy) in the system, it's like turning the dial on that machine to the right: it does more work. Evaporation increases when the temperature rises, and hence arid places grow drier. We call this phenomenon drought, and now we see it everywhere. Cape Town, among the most beautiful cities on earth, spent 2018 flirting with going completely dry. Its four million residents were rationed twenty-three gallons per person per day, enough for a shower, as long as you didn't want to take a drink or flush the toilet. Why? Because of a three-year drought that scientists said, based on past history, should be expected about once a millennium.[14] But of course the phrase "based on past history" no longer makes sense, because that history took place on what was essentially a different planet with a different atmospheric chemistry.

That's why there are versions of the Cape Town story on every continent. A couple of years earlier it was São Paulo, home to twenty million Brazilians, that was turning off the taps. Bangalore may be the highest-tech city in the developing world, with nearly two million IT professionals, but it's also faced drought every year since 2012.[15] The Po River Valley is Italy's agricultural heartland, supplying 35 percent of its crops, but its average temperature is almost four degrees Fahrenheit higher than it was in 1960, and its rainfall has fallen by a fifth. So, by the summer of 2017, an enormous drought forced mayors and governors there to start rationing water. "The Po Plain used to be extraordinarily water-rich, and hence we got used to a situation where water has always been available," said one local official.[16] Most of Italy was affected—Rome shut off its network of public drinking fountains, the largest in the world, and the Vatican turned off the water in the Baroque fountains of St. Peter's Square. But none of it was enough—by September, the source of the Po, on Monviso, in the Cottian Alps, was dry.[17] Petrarch talked about the source of the Po, and so did Chaucer and Dante. But they lived on a planet with 40 percent less carbon dioxide.

As land dries out, it often burns. Humans have converted more and more forest into farmland, which reduces the number of fires overall,[18] but where there's something to combust, fire has become a menace of a different kind. Jerry Williams, the former head firefighter for the U.S.

Forest Service, told a conference not long ago that "my first experience with a really unimaginable fire was in Northern California late in August in 1987," when a thousand blazes broke out simultaneously. "I remember saying, 'Jesus, we will never see anything like that again.' And the next year we saw Yellowstone." Now, he said, "it seems like every year we see a 'worst' one. And the next year we see a worse one yet. They're unbounded."[19] As Michael Kodas reports in his recent book, *Megafire*, fire season is on average seventy-eight days longer across the American West than it was in 1970, and in some parts, it essentially never ends; since 2000, more than a dozen U.S. states have reported the largest wildfires in their recorded histories.[20] We know about those fires because there are reporters nearby, and urban populations to smell the smoke, but there are also now much vaster blazes virtually every spring and summer across Siberia, which we can track only with satellite photos. In fact, by this point there's an obvious rhythm to the global danger: prolonged drought, then a record heat wave, then a spark. Australia's McArthur Forest Fire Danger Index used to top out at 100, but in 2009, after a month of record heat and the lowest rainfalls ever measured, the index reached 165, and 173 people died in a blaze that raced through the suburbs.[21] In 2016, the city at the heart of Alberta's tar sands complex, Fort McMurray, had to be entirely evacuated after a low snowpack gave way to a record spring heat wave and, soon, a May blaze that spread to a million and a half acres, chasing 88,000 people from their homes.[22] In 2018, 80 people died in Attica, in the heart of classical Greece, when a firestorm took off amid record heat; those who survived did so only by diving into the Aegean Sea, even as "flames burned their backs." Two dozen people who couldn't make it to the beach just formed a circle and embraced one another as they died.[23]

Sometimes humans start the fires—sparks from golf clubs hitting rocks have set off several Southern California blazes, and in Utah, target shooters managed to ignite twenty blazes during the drought of 2012.[24] But in a deeper sense, humans help start all of them: each degree Fahrenheit we warm the planet increases the number of lightning strikes by 7 percent,[25] and once fires get going in our hot, dry new world, they are all but impossible to fight. These blazes "make up a new category of fire," Kodas writes, "exhibiting behaviors rarely seen by foresters or firefighters. The infernos can launch fusillades of firebrands miles ahead of the conflagration to ignite new blazes in unburnt forests and communities. The flames create their own weather systems, spinning tornadoes of fire into the air, filling the sky with pyrocumulus clouds that blast the ground with lightning to start new fires, and driving back firefighting aircraft with their winds." They "cannot be controlled by any suppression resources that we have available anywhere in the world," said one Australian researcher.[26]

And the devastation they leave behind—well, you've seen the rows of burned-out houses on your Facebook feed. But imagine all the other effects. In the spring of 2017, after the obligatory deep drought and record heat, Kansas saw the largest wildfire in its history. There weren't many houses in the way, but there was lots and lots of barbed-wire fencing, and all those wooden posts burned to stumps. New fence costs ten thousand dollars a mile, and at many ranches, that alone meant two million dollars or more in uninsured losses. Far worse were the cattle: At a ranch outside Ashland, "dozens of Angus cows lay dead on the blackened ground, hooves jutting in the air. Others staggered around like broken toys, unable to see or breathe, their black fur and dark eyes burned, plastic identification tags melted to their ears," the *New York Times* reported. A sixty-nine-year-old rancher walked among them with a rifle. "They're gentle," he said. "They know us. We know them. You just thought, 'Wow, I am sorry.' You think you're done and the next day you got to go shoot more."[27]

That global pump I've described doesn't just suck water up; it also spews it back out. An easy rule of thumb is that for every drought, a flood. Occasionally they're in the same places a few months apart, but another rule of thumb: dry places get dryer, and wet places wetter.

So: ocean temperatures had risen about a degree Fahrenheit off the Texas coast in recent years, which means, on average, about 3 to 5 percent more water in the atmosphere.[28] And when Hurricane Harvey wandered across the Gulf in August 2017, it crossed a particularly warm and deep eddy, intensifying "at near record pace" into a Category 4 storm. But it wasn't its winds that tied it with Katrina as the most economically damaging storm in American history; it was the rain, which came down in buckets. Not in buckets—in football stadiums. Thirty-four trillion gallons, enough to fill 26,000 New Orleans Superdomes. That's 127 billion tons, enough weight that Houston actually sank by a couple of centimeters. In places, the rainfall topped fifty-four inches, by far the largest rainstorm in American history. "Harvey's rainfall in Houston was 'biblical' in the sense that it likely occurred around once since the Old Testament was written," one study concluded.[29] Because we've warmed the atmosphere, the odds of a storm that could drop that much rain on Texas have gone up sixfold in the last twenty-five years.[30] Three months after the storm, another study found that the rainfall was as much as 40 percent higher than it would have been from a similar storm before we'd spiked the carbon dioxide levels in the atmosphere.[31] When Hurricane Florence hit the Carolinas in September 2018, it set a new record for East Coast rainfall—the storm dumped the equivalent of all the water in Chesapeake Bay.[32]

This isn't something that happens just in Houston. In Calcutta, home to fourteen million people, none of whom is an oil baron and a third of whom reside in flood-prone slums, the number of "cloudburst days" has tripled in the last five decades. "This is what we say to God," one pavement-dwelling mother of four explained. "If a storm comes, kill us and our children at once so no one will be left to suffer."[33] In the Northeast United States, where I live in landlocked Vermont, we've watched extreme precipitation (two inches or more of rain in twenty-four hours) grow 53 percent more common since 1996.[34] (Since *1996*, when the first flip phone was sold.) All that water cascades over all that we've built these last few centuries—a 2018 *New York Times* survey showed that 2,500 of America's toxic chemical sites lie in flood-prone areas.[35] Harvey, for instance, swamped a factory that spilled huge quantities of lye. In effect, we've put the planet on a treadmill, and we keep pushing up the speed. We're used to the idea that geologic history unfolds over boundless eons at a glacial pace, but not when you're changing the rules.

Actually, perhaps it *is* proceeding at a glacial pace; it's just that "glacial" means something different now. All those Hiroshimas' worth of heat are thawing ice at astonishing speed. Much of the sea ice that filled the Arctic in the early pictures from space is gone now—viewed from a distance, Earth looks strikingly different. Everything frozen is melting. A few years ago, the mountaineer and filmmaker David Breashears took his camera into the Himalayas to retake the first images sent home from the roof of the world, during the Mallory expedition of 1924. He spent days climbing to the same crags, and catching the same glaciers from the same angles. Only, now they were hundreds of vertical feet smaller—a Statue of Liberty shorter. And once ice starts to thaw, it's hard to slow down the process. A 2018 study concluded that even if we stopped emitting all greenhouse gases today, more than a third of the planet's glacial ice would melt anyway in the coming decades.[36]

For the moment, though, don't think about the future. Just think about what we've done so far, in the early stages of this massive transformation. Climate change is currently costing the U.S. economy about $240 billion a year,[37] and the world, $1.2 trillion annually, wiping 1.6 percent each year from the planet's GDP.[38] That's not much yet—we're rich enough as a planet that it doesn't profoundly change the overall game—but look at particular places: Puerto Rico, say, after Hurricane Maria ripped it from stem to stern with Category 5 winds. It was the worst natural disaster in a century in America—in the spring of 2018, a Harvard study estimated it had killed nearly five thousand people, twice the number who died in Katrina[39]—and the economic toll guaranteed it would go on stunting lives for years: the total cost was north of $90 billion, for an island whose pre-storm GDP was $100 billion a year. Economists calculated that it would take twenty-six years for the island's economy to get back to where it had been the day before the storm hit[40]—if, of course, another hurricane didn't strike in the meantime.

Or look at people living so close to the margin that small changes make a huge difference. I noted earlier that we've seen a steady decline in extreme poverty and hunger. "Our problem is not too few calories but too many," Steven Pinker wrote smugly.[41] But late in 2017, a UN agency announced that after a decade of decline, the number of chronically malnourished human beings had started growing again, by 38 million, to a total of 815 million, "largely due to the proliferation of violent conflicts and climate-related shocks."[42] In June 2018, researchers said the same sad thing about child labor: after years of decrease, it, too, was on the rise, with 152 million kids at work, "driven by an increase in conflicts and climate induced disasters."[43]

Those "conflicts," too, are ever more closely linked to the damage we've done to the climate. By now, it's a commonplace that record drought helped destabilize Syria, sparking the conflict that sent a million refugees sprawling across Europe and helped poison the politics of the West. (And a 2018 World Bank study predicted that further climate change would displace as many as *143 million* people from Africa, South Asia, and Latin America by 2050. The authors whimsically urged cities to "prepare infrastructure, social services, and employment opportunities ahead of the influx.")[44] But there are a hundred smaller examples. On top of Mount Kenya, two-thirds of the ice cover has disappeared; ten of the eighteen glaciers that once watered the surrounding region are gone altogether. Herders, whose pastures are turning to dust, have started driving their cattle into the farmland nearer the mountain. "Our cows had nothing to eat," explained one man. "Would you let your cow die if there is grass somewhere near?" The farmers who till that land (traditionally from different ethnic groups) have fought back hard, and people have died. "I have not slept for two days," one farmer said. "If I do, they will bring their cows and let them loose in our farms. They are lurking, waiting for us to sleep, then bring their cows and goats to eat our cabbages and maize."[45] There are studies that try to quantify these changes—one standard deviation increase in temperature supposedly increases conflicts between groups by 14 percent[46]—but you hardly need them. Common sense will do. The planet is crowded. As we begin to change it, people are pushed closer together. We know what happens next.

There was hope, thirty years ago, that global warming might somehow limit itself, that raising the temperature might trigger some other change that would cool the planet. Clouds, perhaps: as the atmosphere grew moister with increased evaporation, more clouds might form, blocking some of the incoming sunlight. No such luck; if anything, the kinds of clouds we're producing on a hotter planet seem to be trapping more heat and making it hotter still.[47]

Such feedback loops, it turns out, lie buried in all kinds of earth systems, and so far, they're all making the problem worse, not better. When the white ice melts in the Arctic, it stops reflecting the sun's rays back out to space: a shiny mirror is replaced with dull blue seawater, which absorbs the sun's heat. The sea surface temperature has gone up by seven degrees Fahrenheit in recent years in parts of the Arctic.[48] Hidden ice, locked beneath the soils of the Arctic, is now starting to melt fast, too, and as that permafrost thaws, microbes convert some of the frozen organic material into methane and carbon dioxide, which cause yet more warming—perhaps, say scientists, enough to add a degree and a half Fahrenheit or more to the eventual warming.[49]

New studies also show that degradation of tropical woodlands, from wildfire, drought, and selective logging, has turned them from sinks for carbon into sources of more carbon dioxide. This transition is important. When economists scoffed at books such as *Limits to Growth*, insisting that scarcity, and the resulting higher prices, would spur the search for new sources, they had a point: we haven't run out of copper; and oil obviously keeps flowing. But places to put our waste? Those are ever harder to come by, as the increasing temperature weakens the ability of forests and oceans to soak up carbon. Should this weakening continue, the *New York Times* noted, "the result would be something akin to garbage workers going on strike, but on a grand scale: The amount of carbon dioxide in the atmosphere would rise faster, speeding global warming even beyond its present rate."[50] And that's what seems to be happening. Even as our emissions rise more slowly, the amount of carbon dioxide in the atmosphere keeps spiking faster.

But, again, we're getting ahead of the story. Right now, just focus on what we've already done, how much we've already changed our world. Consider California, the Golden State, long the idyllic picture of the human future. It endured a horrific five-year drought at the start of this decade, the deepest in thousands of years—so deep that the state was tapping into groundwater that was twenty thousand years old, rain that fell during the last Ice Age;[51] so deep that the state's Sierra Nevada range rose an inch just because sixty-three trillion gallons of water had evaporated;[52] so deep that it killed 102 million trees, a blight "unprecedented in our modern history," in the words of the *Los Angeles Times*. (Sugar pines should live five hundred years, but "everywhere you walk, through certain parts of the forest, half these big guys are dead," said one forester.[53]) The drought ended in the winter of 2017, when the rains finally came, an endless atmospheric river that poured off the hot Pacific into the high mountains. Everyone breathed a sigh. California's authorities said, of course, that they understood the reprieve was only temporary, but they could be forgiven for relaxing a little as the hills turned lush and green. (Anyway, they were having enough trouble with the floods that the record rainfall produced: the deluge caused almost a billion dollars in damage, for instance, to the nation's highest dam.)

The summer of 2017, though, proved hotter and drier than even in the worst years of the drought, and all that green grass soon browned up, and in October, a firestorm swept through Napa and Sonoma. Despite all the TV alerts and text warnings, it killed more people in a shorter time than any American fire in a century—old people, especially, who simply couldn't outrun the flames. Reporters described the "apocalyptic scene" in neighborhoods that, twenty-four hours earlier, came as close to representing the good life as any place on earth. People died in swimming pools where they'd taken shelter, the tiled sides as "hot as oven racks." In Santa Rosa, "the aluminum wheels on cars melted and dripped down driveways like tiny rivers of mercury before hardening. A pile of bottles melded together into a tangle so contorted it looked like a Picasso sculpture. Plastic garbage bins were reduced to mere stains on the pavement."[54]

But that was October, right at the end of the state's traditional fire season. So, people breathed a sigh again, and began the work of cleaning up, knowing they had a little time—until, in December, record heat and dryness in Southern California touched off what became the largest blaze in California history, a blaze that "jumped ten-lane highways with unsettling ease"[55] and threatened the homes of Rupert Murdoch, Elon Musk, and Beyoncé.

It burned till the New Year, and then, as 2018 began, people breathed a sigh again, and welcomed the prospect of a little winter rain. The first storm hit, and it brought prodigious amounts of water, half an inch inside of five minutes in some places. That water fell on the burned-over hills, where there were no plants left to hold the flood, and so it turned into a mudslide. Twenty-one people died. Those who survived remembered the sound, a rumbling like a freight train. "It buried houses and cars and people," said the writer Nora Gallagher, who lived nearby. "It buried the freeway and the train tracks. The creeks ran black with ash. It went all the way to the ocean. A body of a man was found there on the beach. Not far away from him was the body of a bear."[56]

The rest of 2018 was no better. By early August, the record for the largest fire in state history had fallen again, to a massive blaze in the Mendocino area; Yosemite Valley was closed "indefinitely" as flames licked at its access roads; and meteorologists were trying to make sense of a vast "fire tornado" that rose 39,000 feet above the city of Redding and twisted so violently it stripped the bark from trees. And then, in autumn, came the most gruesome fire of all, in the Sierra foothills above Chico. After the fall "rainy season" delivered a seventh of an inch of precipitation, the town of Paradise exploded in flames; people died by the score in their cars as they tried to flee down narrow, burning roads. The president blamed the conflagration on "forest mismanagement" and recommended "raking"; meanwhile forensic teams tried to recover the victims' DNA from the ashes of burned-out subdivisions.

This is our reality right now. It will get worse, but it's already very, very bad. Nora Gallagher again: "Climate believers, climate deniers, deep in our hearts we think it will happen somewhere else. In some other place—we don't actually say this but we may think it—in a poorer one, say, Puerto Rico or New Orleans or Cape Town or one of those islands where the sea level is rising. Or it will happen in some other time, in 2025 or 2040 or next year. But we are here to tell you, in this postcard from the former paradise, that it won't happen next year, or somewhere else. It will happen right where you live and it could happen today. No one will be spared."[57]

3

Oh, it could get *very* bad.

In 2015 a study in the *Journal of Mathematical Biology* pointed out that if the world's oceans kept warming, by 2100 they might become hot enough to "stop oxygen production by phytoplankton by disrupting the process of photosynthesis." Given that two-thirds of the earth's oxygen comes from phytoplankton, that would "likely result in the mass mortality of animals and humans."[1]

A year later, above the Arctic Circle, in Siberia, a heat wave thawed a reindeer carcass that had been trapped in the permafrost. The exposed body released anthrax into nearby water and soil, infecting two thousand reindeer grazing nearby, and they in turn infected some humans; a twelve-year-old boy died. As it turns out, permafrost is a "very good preserver of microbes and viruses, because it is cold, there is no oxygen, and it is dark"—scientists have managed to revive an eight-million-year-old bacterium they found beneath the surface of a glacier. Researchers believe there are fragments of the Spanish flu virus, smallpox, and bubonic plague buried in Siberia and Alaska.[2]

Or consider this: as ice sheets melt, they take weight off land, and that can trigger earthquakes—seismic activity is already increasing in Greenland and Alaska. Meanwhile, the added weight of the new seawater starts to bend the earth's crust. "That will give you a massive increase in volcanic activity. It'll activate faults to create earthquakes, submarine landslides, tsunamis, the whole lot," explained the director of University College London's Hazard Centre.[3] Such a landslide happened in Scandinavia about eight thousand years ago, as the last Ice Age retreated and a Kentucky-size section of Norway's continental shelf gave way, "plummeting down to the abyssal plain and creating a series of titanic waves that roared forth with a vengeance," wiping all signs of life from coastal Norway to Greenland and "drowning the Wales-sized landmass that once connected Britain to the Netherlands, Denmark, and Germany." When the waves hit the Shetlands, they were sixty-five feet high.[4]

There's even this: if we keep raising carbon dioxide levels, we may not be able to think straight anymore. At a thousand parts per million (which is within the realm of possibility for 2100), human cognitive ability falls 21 percent. "The largest effects were seen for Crisis Response, Information Usage, and Strategy," a Harvard study reported, which is too bad, as those skills are what we seem to need most.[5]

I could, in other words, do my best to scare you silly. I'm not opposed on principle—changing something as fundamental as the composition of the atmosphere, and hence the heat balance of planet, is certain to trigger all manner of horror, and we shouldn't shy away from it. The dramatic uncertainty that lies ahead may be the most frightening development of all; the physical world is going from backdrop to foreground. (It's like the contrast between politics in the old days, when you could forget about Washington for weeks at a time, and politics in the Trump era, when the president is always jumping out from behind a tree to yell at you.)

But just as we were careful earlier to stick with that which had already happened, now that we're considering the future, let's try to occupy ourselves with the most likely scenarios, because they are more than disturbing enough. Long before we get to tidal waves or smallpox, long before we choke to death or stop thinking clearly, we will need to concentrate on the most mundane and basic facts: everyone needs to eat every day, and an awful lot of us live near the ocean.

Food supply first. We've had an amazing run since the end of World War II, with crop yields growing fast enough to keep ahead of a fast-rising population. It's come at great human cost—displaced peasant farmers fill many of the planet's vast slums—but in terms of sheer volume, the Green Revolution's fertilizers, pesticides, and machinery managed to push output sharply upward. That climb, however, now seems to be running into the brute facts of heat and drought. There are studies to demonstrate the dire effects of warming on coffee, cacao, chickpeas, and champagne, but it is cereals that we really need to worry about, given that they supply most of the planet's calories: corn, wheat, and rice all evolved as crops in the climate of the last ten thousand years, and though plant breeders can change them, there are limits to those changes. You can move a person from Hanoi to Edmonton, and she might decide to open a Vietnamese restaurant. But if you move a rice plant, it will die.

A 2017 study in Australia, home to some of the world's highest-tech farming, found that "wheat productivity has flatlined as a direct result of climate change." After tripling between 1900 and 1990, wheat yields had stagnated since, as temperatures increased a degree and rainfall declined by nearly a third. "The chance of that just being variable climate without the underlying factor [of climate change] is less than one in a hundred billion," the researchers said, and it meant that despite all the expensive new technology farmers kept introducing, "they have succeeded only in standing still, not in moving forward." Assuming the same trends continued, yields would actually start to decline inside of two decades, they reported.[6] In June 2018, researchers found that a two-degree Celsius rise in temperature—which, recall, is what the Paris accords are now *aiming for*—could cut U.S. corn yields by 18 percent.

A four-degree increase—which is where our current trajectory will take us—would cut the crop almost in half. The United States is the world's largest producer of corn, which in turn is the planet's most widely grown crop.[7]

Corn is vulnerable because even a week of high temperatures at the key moment can keep it from fertilizing. ("You only get one chance to pollinate a quadrillion kernels of corn," the head of a commodity consulting firm explained.)[8] But even the hardiest crops are susceptible. Sorghum, for instance, which is a staple for half a billion humans, is particularly hardy in dry conditions because it has big, fibrous roots that reach far down into the earth. Even it has limits, though, and they are being reached. Thirty years of data from the American Midwest show that heat waves affect the "vapor pressure deficit," the difference between the water vapor in the sorghum leaf's interior and that in the surrounding air. Hotter weather means the sorghum releases more moisture into the atmosphere. Warm the planet's temperature by two degrees Celsius—which is, again, now the world's *goal*—and sorghum yields drop 17 percent. Warm it five degrees Celsius (nine degrees Fahrenheit), and yields drop almost 60 percent.[9]

Except perhaps for asphalt shingles, it's hard to imagine a topic duller than sorghum yields. It's the precise opposite of clickbait. But people have to eat; in the human game, the single most important question is probably "What's for dinner?" And when the answer is "Not much," things deteriorate fast. In 2010 a severe heat wave hit Russia, and it wrecked the grain harvest, which led the Kremlin to ban exports. The global price of wheat spiked, and that helped trigger the Arab Spring—Egypt at the time was the largest wheat importer on the planet. That experience set academics and insurers to work gaming out what the next food shock might look like. In 2017 one team imagined a vigorous El Niño, with the attendant floods and droughts—for a season, in their scenario, corn and soy yields declined by 10 percent, and wheat and rice by 7 percent. The result was chaos: "quadrupled commodity prices, civil unrest, significant negative humanitarian consequences . . . Food riots break out in urban areas across the Middle East, North Africa, and Latin America. The euro weakens and the main European stock markets lose ten percent of their value."[10]

At about the same time, a team of British researchers released a study demonstrating that even if you can grow plenty of food, the transportation system that distributes it runs through just fourteen major chokepoints, and those are vulnerable to—you guessed it—massive disruption from climate change. For instance, U.S. rivers and canals carry a third of the world's corn and soy, and they've been frequently shut down or crimped by flooding and drought in recent years. Brazil accounts for 17 percent of the world's grain exports, but heavy rainfall in 2017 stranded three thousand trucks. "It's the glide path to a perfect storm," said one of the report's authors.[11]

Five weeks after *that*, another report raised an even deeper question. What if you can figure out how to grow plenty of food, and you can figure out how to guarantee its distribution, but the food itself has lost much of its value? The paper, in the journal *Environmental Research*, said that rising carbon dioxide levels, by speeding plant growth, seem to have reduced the amount of protein in basic staple crops, a finding so startling that, for many years, agronomists had overlooked hints that it was happening. But it seems to be true: when researchers grow grain at the carbon dioxide levels we expect for later this century, they find that minerals such as calcium and iron drop by 8 percent, and protein by about the same amount. In the developing world, where people rely on plants for their protein, that means huge reductions in nutrition: India alone could lose 5 percent of the protein in its total diet, putting 53 million people at new risk for protein deficiency. The loss of zinc, essential for maternal and infant health, could endanger 138 million people around the world.[12] In 2018, rice researchers found "significantly less protein" when they grew eighteen varieties of rice in high–carbon dioxide test plots. "The idea that food became less nutritious was a surprise," said one researcher. "It's not intuitive. But I think we should continue to expect surprises. We are completely altering the biophysical conditions that underpin our food system."[13] And not just ours. People don't depend on goldenrod, for instance, but bees do. When scientists looked at samples of goldenrod in the Smithsonian that dated back to 1842, they found that the protein content of its pollen had "declined by a third since the industrial revolution—and the change closely tracks with the rise in carbon dioxide."[14]

Bees help crops, obviously, so that's scary news. But in August 2018, a massive new study found something just as frightening: crop pests were thriving in the new heat. "It gets better and better for them," said one University of Colorado researcher. Even if we hit the UN target of limiting temperature rise to two degrees Celsius, pests should cut wheat yields by 46 percent, corn by 31 percent, and rice by 19 percent. "Warmer temperatures accelerate the metabolism of insect pests like aphids and corn borers at a predictable rate," the researchers found. "That makes them hungrier[,] and warmer temperatures also speed up their reproduction." Even fossilized plants from fifty million years ago make the point: "Plant damage from insects correlated with rising and falling temperatures, reaching a maximum during the warmest periods."[15]

Just as people have gotten used to eating a certain amount of food every day, they've gotten used to living in particular places. For obvious reasons, many of these places are right by the ocean: estuaries, where rivers meet the sea, are among the richest ecosystems on earth, and water makes for easy trade. From the earliest cities (Athens, Corinth, Rhodes) to the biggest modern metropolises (Shanghai, New York, Mumbai), proximity to saltwater meant wealth and power. And now it means exquisite, likely fatal, vulnerability.

Throughout the Holocene (the ten-thousand-year period that began as the last ice age ceased, the stretch that encompasses all recorded human history), the carbon dioxide level in the atmosphere stayed stable, and therefore so did the sea level, and hence it took a while for people to worry about sea level rise. The United Nations' Intergovernmental

Panel on Climate Change (IPCC) predicted in 2003 that sea level should rise a mere half meter by the end of the twenty-first century, most of that coming because warm water takes up more space than cold, and while a half meter would be enough to cause expense and trouble, it wouldn't really interfere with settlement patterns.[16] But even as the IPCC scientists made that estimate, they cautioned that it didn't take into account the possible melt of the great ice sheets over Greenland and Antarctica. And pretty much everything we've learned in the years since makes scientists think that those ice sheets are horribly vulnerable.

Paleoclimatologists, for instance, have discovered that in the distant past, sea levels often rose and fell with breathtaking speed. Fourteen thousand years ago, as the Ice Age began to loosen its grip, huge amounts of ice thawed in what researchers call meltwater pulse 1A, raising the sea level by sixty feet.[17] Thirteen feet of that may have come in a single century. Another team found that millions of years ago, during the Pliocene, with carbon dioxide levels about where they are now, the West Antarctic Ice Sheet seems to have collapsed in as little as a hundred years.[18] "The latest field data out of West Antarctica is kind of an OMG thing," a federal official said in 2016—and that was before the *really* epochal news in the early summer of 2018, when eighty-four researchers from forty-four institutions pooled their data and concluded that the frozen continent had lost three trillion tons of ice in the last three decades, with the rate of melt *tripling* since 2012.[19] As a result, scientists are now revising their estimates steadily upward. Not half a meter of sea level rise, but a meter. Or two meters. "Several meters in the next fifty to 150 years," said James Hansen, the planet's premier climatologist, who added that such a rise would make coastal cities "practically ungovernable."[20] As Jeff Goodell (who in 2017 wrote the most comprehensive book to date on sea level rise) put it, such a rise would "create generations of climate refugees that will make today's Syrian war refugee crisis look like a high school drama production."[21]

What's really breathtaking is how ill-prepared we are for such changes. Goodell spent months reporting in Miami Beach, which was literally built on sand dredged up from the bottom of Biscayne Bay. He managed to track down Florida's biggest developer, Jorge Pérez, at a museum opening. Pérez was not, he insisted, worried about the rising sea because "I believe that in twenty or thirty years, someone is going to find a solution for this. If it is a problem for Miami, it will also be a problem for New York and Boston—so where are people going to go?" (He added, with Trump-level narcissism, "Besides, by that time I'll be dead, so what does it matter?")[22] To the extent that we're planning at all, it's for the old, low predictions of a meter or less. Venice, for instance, is spending $6 billion on a series of inflatable booms to hold back storm tides. But they're designed to stop sea level rise of about a foot. New York City is building a "U-Barrier," a berm to protect Lower Manhattan from inundation in a storm the size of Hurricane Sandy. But as the sea level rises, winds like Sandy's will drive far more water into Manhattan, so why not build it higher? "Because the cost goes up exponentially," said the architect.[23] The cost is already starting to mount. Researchers showed in 2018 that Florida homes near the flood lines were selling at a 7 percent discount, a figure growing over time because "sophisticated buyers" know what is coming.[24] Insurance companies are balking: basements from "New York to Mumbai" may be uninsurable by 2020, the CEO of one of Europe's largest insurers said in 2018.[25]

Some of the cost of climate change can be measured in units we're used to dealing with. Testimony submitted by climate scientists to a federal court in 2017, for instance, said that if we don't take much stronger action now, future citizens would have to pay $535 trillion to cope with global warming.[26] How is that possible? Take one small county in Florida, which needs to raise 150 miles of road to prevent flooding from even minimal sea level rise. That costs $7 million a mile, putting the price tag at over $1 billion, in a county that has an annual road budget of $25 million. Or consider the numbers from Alaska, where officials are preparing to move one coastal village with four hundred residents that's threatened by rising waters at a cost of up to $400 million—$1 million a person.[27] Multiply this by everyone everywhere, and you understand why the costs run so high. A team of economists predicted a 12 percent risk that global warming could reduce global economic output by 50 percent by 2100—that is to say, there's a one-in-eight chance of something eight times as bad as the Great Recession.[28]

But some things can't be measured, and the damage there seems even greater. For instance, the median estimate, from the International Organization for Migration, is that we may see two hundred million climate refugees by 2050. (The high estimate is a billion.) Already "the likelihood of being uprooted from one's home has increased sixty percent compared with forty years ago."[29] The U.S. military frets about that because masses of people on the march destabilize entire regions. "Security will start to crumble pretty quickly," said Adm. Samuel Locklear, former chief of U.S. Pacific Command, explaining why climate change was his single greatest worry.[30]

The biggest worry for people losing their homes is . . . losing their homes. So, let me tell you about a trip I took last summer, to the ice shelf of Greenland. I was with a pair of veteran ice scientists and two young poets—a woman named Kathy Jetnil-Kijiner, from the Marshall Islands in the Pacific, and another named Aka Niviana, who was born on this largest of all the earth's islands, a massive sheet of ice that, when it melts, will raise the level of the oceans more than twenty feet.

And it is melting. We landed at the World War II–era airstrip in Narsarsuaq and proceeded by boat through the iceberg-clogged Tunulliarfik Fjord, arriving eventually at the foot of the Qaterlait Glacier. We hauled gear up the sloping, icy ramp of the glacier and made camp on an outcrop of red granite bedrock nearly a kilometer inland. In fact, we made camp twice, because the afternoon sun swelled the stream we'd chosen for a site, and soon the tents were inundated. But after dinner, in the late Arctic sunlight, the two women donned the traditional dress of their respective homelands and hiked farther up the glacier, till they could see both the ocean and the high ice. And there they performed a poem they'd composed, a cry from angry and engaged hearts about the overwhelming fact of their lives.

The ice of Niviana's homeland was disappearing, and with it a way of life. While we were on the ice sheet, researchers reported that "the oldest and thickest sea ice" in the Arctic had melted, "opening waters north of Greenland that are normally frozen even in summer."[31] Just up the coast from our camp, a landslide triggered by melting ice had recently set off a hundred-foot tsunami that killed four people in a remote village: it was, said scientists, precisely the kind of event that will "become more frequent as the climate warms."[32]

The effect, however, is likely to be even more immediate on Jetnil-Kijiner's home. The Marshalls are a meter or two above sea level, and already the "king tides" wash through living rooms and unearth graveyards. The breadfruit trees and the banana palms are wilting as saltwater intrudes on the small lens of fresh water that has supported life on the atolls for millennia. Jetnil-Kijiner was literally standing on the ice that, as it melts, will drown her home, leaving her and her countrymen with, as she put it, "only a passport to call home."

So, you can understand the quiet rage that flowed through the poem the two women had written, a poem they now shouted into a chill wind on this glacier that flowed up to the great ice sheet, silhouetted against the hemisphere's starkest landscape. It was a fury that came from a long and bitter history: the Marshalls were the site of the atom bomb tests after the war, and Bikini Atoll remains uninhabitable, just as the United States left nuclear waste lying around the ice when it abandoned the thirty bases it had built in Greenland.

The very same beasts
That now decide
Who should live
And who should die . . .
We demand that the world see beyond
SUVs, ACs, their pre-packaged convenience

Their oil-slicked dreams, beyond the belief
That tomorrow will never happen

But, of course, climate change is different, the first crisis that, though it affects the most vulnerable first and hardest, will eventually come for us all.

Let me bring my home to yours
Let's watch as Miami, New York,
Shanghai, Amsterdam, London
Rio de Janeiro and Osaka
Try to breathe underwater . . .
None of us is immune.

Science can tell us a good deal about this crisis. Jason Box, an American glaciologist who organized the trip, has spent the last twenty-five years journeying to Greenland. "We called this place where we are now the Eagle Glacier because of its shape when we first came here five years ago," Box said. "But now the head and the wings of the bird have melted away. I don't know what we should call it now, but the eagle is dead." He busied himself replacing the batteries in his remote weather stations, scattered across the ice. They tell one story, but his colleague Alun Hubbard, a Welsh scientist, conceded that there were limits to what instruments could explain. "It's just gobsmacking looking at the trauma of the landscape," he said. "I just couldn't register the scale of how the ice sheet had changed in my head."

But artists *can* register scale. They can transpose the fact of melting ice to inundated homes and bewildered lives, gauge it against long history and lost future. Science and economics have no real way to value the fact that people have lived for millennia in a certain rhythm, have eaten the food and sung the songs of certain places that are now disappearing. This is a cost only art can measure, and it makes sense that the units of that measurement are sadness and fury—and also, remarkably, hope. The women's poem, shouted into the chill wind, ended like this:

Life in all forms demands
The same respect we all give to money . . .
So each and every one of us
Has to decide
If we
Will
Rise

And so, we must—in fact, this book will end with a description of what that rising might look like. But if, as now seems certain, the melt continues, then the villages of the Marshalls and the ports of Greenland will be overwhelmed. And we will all be a little poorer, because a way of being will have been cut off. The puzzle of being human will have lost some of its oldest, most artful pieces.

"The loss of Venice," Jeff Goodell writes, wouldn't be about just the loss of present-day Venetians. "It's the loss of the stones in the narrow streets where Titian and Giorgione walked. It's the loss of eleventh-century mosaics in the basilica, and the unburied home of Marco Polo, and palazzos along the Grand Canal. . . . The loss of Venice is about the loss of a part of ourselves that reaches back in time and binds us together as civilized people."[33]

We all have losses already. Where I live, it's the seasons: winter doesn't reliably mean winter anymore, and so the way we've always viscerally told time has begun to break down. In California, it's the sense of ease: the smell of the fire next time lingers in the eucalyptus groves. There are many ways to be poorer, and we're going to find out all of them.

4

This is a book about being human, but for a moment we need to leave people a little behind and go out to the deep ocean and back into deep time—go, that is, to realms so staggering in their size that we can finally understand the scale of our very human impact.

Saltwater first. We erred, though understandably, when we named our planet. "Ocean" would have been more apt, as 70 percent of the surface is covered by the seas. We live on their margins, so our worries are dominated by their rise, their intrusion into our sphere. But unless you earn your livelihood by fishing, you rarely think very much about what goes on beneath the surface. The sea seems so endlessly vast; very few humans ever disappear over the edge of the horizon and get a sense of that blue expanse—which is why, of course, we've mistreated it so casually. By the middle of this century the ocean may contain more plastic than fish by weight,[1] partly because we toss away so many bottles and partly because we take far more life from the ocean than it can reproduce. Since 1950 we've wiped out perhaps 90 percent of the big fish in the ocean: swordfish, marlin, grouper. This is not surprising when one bluefin tuna can bring $180,000 on the Japanese market, or when 270,000 sharks are killed each day, mostly for their fins, which add no taste but much status to bowls of soup. Every year, we plow an undersea area twice the size of the continental United States, with trawlers leveling everything on the sea floor. Were this on the surface, there would be real protest, but it's invisible.[2]

Still, the overfishing, and the dead zones at the mouths of all major rivers where fertilizers pour into the sea, and the gyres of plastic spinning slowly a thousand miles offshore—these are the smallest of our insults to the ocean. The overwhelming threat comes, again, from the fossil fuel we burn and the effects of the carbon dioxide that it produces, effects that are even larger in the sea than on the land. For one, the ocean is where most of that extra heat accumulates. Though we focus on the heat in the air around us, about 93 percent of the extra heat is actually collecting in the sea. The deep sea is now warming about nine times faster than it was in the 1960s, '70s, or '80s.[3] (It goes without saying that the Trump administration has proposed big cutbacks for the agency that maintains the network of temperature-monitoring buoys.) To get a sense of how much heat the oceans absorb, consider this: without them, the temperature of the atmosphere would have gone up by 97 degrees Fahrenheit since 1955.[4]

You needn't be an expert diver to see the most dramatic effects of all that heat. A snorkel will do, or even a lungful of air. From Port Douglas, in Queensland, Australia, it's about two hours by motorboat to the outer edge of the Great Barrier Reef. On my last trip, in the spring of 2018, the sea was choppy and no one talked much, just dozed in the early morning sun. We were headed to the Opal Reef, where, three years earlier, a crew had filmed some of the remarkable scenes of coral spawning for the BBC series *Blue Planet II*. Guided by the phases of the moon, and with David Attenborough providing discreet and tasteful narration, the garden of corals had simultaneously released clouds of eggs and sperm for the cameras, in the world's most profligate display of fecundity, a spectacle if there ever was one. But no longer. We moored, tugged on snorkels and masks, and stepped off the stern, clad in full-body "stinger suits" to protect us from the jellyfish. I swam the fifty yards to the reef with James Kerry, a reef researcher at Queensland's James Cook University, and when we got to the edge, we peered down—and it was like snorkeling over an empty parking garage. The *forms* of coral remained—there were exuberant fans and antlers and trays—but instead of vivid neon colors, there were just shades of murk. The living coral were so few in number that Kerry could tell me about them individually as we treaded water at the surface. "That blue one is *Pocillopora damicornis*. It's pretty hardy," he said. "And did you see that one thing down on the bottom that looked like a pillow? That's a large single-polyp coral, a fungid." Some fish wandered through—mostly parrotfish, which feed off the algae that covers the dead coral.

The Great Barrier Reef is the largest living structure on Earth, but it is roughly half as living as it was three years ago. Massive bleaching events in 2016 and 2017 (caused by the incursions of hot water that are becoming ever more common) devastated the northern and central sections. It's hard to explain how grim it looked beneath the surface, in the same way that it's hard to explain precisely how you know instinctively that a dead body is dead. But everyone on board the motorboat that day was trying to put it in words, working out a kind of grief. Dean Miller, a reef scientist who is the director of media and science for the Great Barrier Reef Legacy, an organization that works to get more scientists out on the water, has filmed transects across this section of the reef over the years, mapping the same route across the coral. "This place—it looked like someone had created the reef, had planted all their favorite corals in perfect shapes and sizes. It was a bustling city, like in *Finding Nemo*. But now it just seems quiet, like the lights have been turned off." The shapes of the dead coral will persist until the first big storm crumbles them, but no one has seen any spawning in the area in the past two years. One reef expert, Jon Brodie, told reporters that the reef was now in a "terminal stage. We've given up. It's been my life," he said. "We've failed."[5] The response of local officials was, of course, emblematic: the head of the Queensland tourism association called the chief coral scientist "a dick," and demanded that the government cut funding for future research. Tourists, he pointed out, "won't do long-haul trips if they think the reef is dead."[6]

The extra heat is not the only thing we're adding to the oceans. As our factories, cars, and furnaces have produced a great plume of carbon dioxide in the atmosphere, some of that carbon dioxide has been absorbed by seawater—at the moment, about a million tons crosses from the air into the sea each hour.[7] For many decades, scientists thought of this as a blessing—some of the carbon that we spewed out would simply disappear into the briny deep—but about fifteen years ago, researchers began to sense profound danger. In the volumes that we are producing carbon, it turns out, even the ocean is too small to soak it up without effect. As carbon dioxide flows into the ocean, some of it is turned into carbonic acid, and that in turn reduces the pH of seawater. Throughout the Holocene—throughout, that is, recorded human history—the ocean's pH was a steady 8.2. It has now dropped to 8.1, which doesn't sound like much, until you remember that pH is a logarithmic scale. The oceans have seen their acidity increase by about 30 percent. At current emission rates, the pH of the oceans will drop to 7.8 or 7.7 by century's end, "well beyond what fish and other marine organisms can tolerate in the laboratory without very serious implications for health, reproduction, and mobility," according to the veteran oceanographer Eelco Rohling.[8] Human bodies, for instance, have a pH of about 7.4. If that drops by 0.2 units (about half what we expect for the ocean this century), it can "cause serious health issues, such as seizures, coma, and death." As water acidifies, the small phytoplankton at the very base of the planet's biological chains struggle to form carbonate for their skeletal parts. Fish blood pH is in equilibrium with the surrounding water; if the water becomes more acid, fish use huge amounts of energy to restore the balance in their cells, which stifles their growth and slows their mobility.[9]

So, carbon dioxide is both heating and acidifying the oceans that cover most of our planet. Taken together, says Rohling, "it's a double whammy" that will "dramatically reduce diversity and numbers of key species throughout the marine ecosystem. This could induce a species collapse from the top of the food web downward."[10] He's not alone in his fears. In 2013, 540 of the world's top ocean scientists collaborated on a report for the United Nations. Their prediction: the oceans, over the course of the century, would become "hot, sour, and breathless."[11] Scientists now routinely predict that by 2050, virtually all the world's coral reefs will be dead.[12] That's roughly thirty years away. That is to say, the same distance in the future as the collapse of the Soviet Union is in the past. It's half as distant in time as the invention of the Frisbee or the birth of Donny Osmond. A child born today who wants to become a marine biologist will barely have her PhD by 2050.

Politicians who don't wish to deal with the issue of global warming often say, "The climate is always changing." Even those people who are properly worried often default to a truism of their own: "The earth will be fine; it's humans who are in trouble." Both statements are technically accurate: no system is perfectly stable, and until the sun blows up some billions of years hence, there will always be a rock orbiting at a distance of about ninety million miles. But both statements are at their core quite wrong: the climate change we are currently forcing will be enormous in comparison with anything our civilization has ever known, and it will fundamentally degrade the earth's biology. Human beings are now a geological force. In fact, we are one of the half-dozen or so largest geological forces to punctuate the billions of years of earth's history.

We know about the previous great disruptions in the Earth's biology because of the fossil record. Five hundred forty million years ago most of the major animal phyla appeared in the fossil record. We call this the "Cambrian explosion," using the word *explosion* in its joyful sense. Life seemed suddenly abundant, as evolution tried out many schemes for thriving on our planet. Five times since, much of that life has suddenly disappeared. We call these periods mass extinctions. The common thread that runs through them all is carbon dioxide.

It's hard to look back 443 million years to the end of the Ordovician and the first of those debacles, but clearly something went "haywire" in the carbon cycle, which shows "wild swings" throughout the "catastrophe," in the assessment of Peter Brannen, in his fine book on extinction.[13] Brannen quotes one geologist who specializes in the period: "When there are severe, rapid changes in the carbon cycle it doesn't end well."[14] We know a good deal more about the Permian extinction, 250 million years ago. Among other things, we know it was the worst of all time, and almost the end of all life on earth. The cause was volcanism—not the eruption of charismatic cinder cones like Mount Fuji, but "burbling floods of lava" that poured out of formations called the Siberian Traps. Volcanoes produce lots of carbon dioxide themselves, but in this case the lava also lit off huge deposits of coal, oil, and gas that had built up over hundreds of millions of years.[15] Before long (in geologic time), the earth was a kind of hell, the ocean was profoundly acidified, and most of the world's species were gone forever—this is the only extinction that did serious damage even to the insect kingdom.

A similar "continental flood basalt," though this time along fissures from Long Island to Quebec and Mauritania to Morocco, triggered the Triassic-Jurassic extinction, which cleared the planet for the dinosaurs to flourish. And then, sixty-five million years ago, came the event that wiped those dinosaurs out. The end of the Cretaceous is the moment most of us think about when we think about extinctions, and what we see in our mind's eye is a giant asteroid hurtling in from outer space, "a rock larger than Mt. Everest traveling twenty times faster than a bullet" (in Peter Brannen's description), one that carved a giant crater into the Gulf of Mexico, triggering a tsunami a thousand feet high and sending huge quantities of earth up into space, which then returned in a "worldwide blizzard of meteorites."[16] Cue the exit of lumbering Tyrannosaurus and the eventual rise of . . . us.

Except that the Hollywood picture of the end of the Cretaceous turns out to be, if not wrong, then considerably more complicated. Something else slightly less dramatic but at least as big was going on across the planet from the asteroid strike: the eruption of yet another massive continental flood basalt, this time in what are called the Deccan Traps, in present-day India. "So profound was this Indian volcanism that it would have been enough to cover the entire lower forty-eight United States in 600 feet of lava," according to Brannen. And enough to do most of the work of driving the fifth great mass extinction via the usual route: carbon dioxide, global warming, ocean acidification. It's possible that the asteroid "was the gun and the Deccan Traps the bullet."[17] The volcanic eruptions were already under way when the asteroid hit, but studies released in 2018 indicate that its impact "fueled an acceleration,"[18] perhaps opening new fissures underwater along the edges of the tectonic plates.[19]

In our time, another cloud of carbon dioxide once again envelops the planet. This time it's not coming from volcanoes; it's coming from tailpipes and smokestacks. Continent-size floods of lava are not setting vast deposits of coal on fire; instead, continent-size power grids are burning through vast deposits of coal. V-8 engines work as effectively as volcanoes, it turns out—and, surprisingly, a good deal faster.

There's not as much carbon to burn as there was at the end of the Permian—carbon dioxide concentrations in the atmosphere will never reach peaks anywhere near as high—but we are burning our hydrocarbons far more quickly than those "continental basalt floods." For two hundred years, human economic activity has largely consisted of digging up fossil fuels and setting them alight, and while two hundred years seems like a long time to us, in geological terms it's like a bat out of . . . well, out of hell. We're currently injecting carbon dioxide into the atmosphere ten times faster than during the End-Permian, which was, just to repeat, the worst event in the earth's history.[20] (If you compare it to the extinction crisis at the end of the Devonian, 360 million years ago, we're pushing carbon dioxide into the atmosphere at somewhere between 12,000 and 40,000 times the ancient rate.)[21] During the End-Permian, which wiped out 90 percent of marine species, the ocean acidified by 0.7 pH units over the course of 10,000 years; on current trends, we will have dropped the pH by 0.5 units in the 250 years ending in 2100.[22] We emit 40 gigatons of carbon dioxide annually at the moment. Our leaders express pride that we seem to be plateauing around that level, but that level is the fastest rate at any time in the last 300 million years, which contains the End-Permian—which, remember, was the definition of bad. Seth Burgess of the U.S. Geological Survey recently published new research on the pulse of carbon dioxide that came as those ancient Siberian lava flows burned all that coal. A reporter asked him if it was appropriate to compare the event with our current situation. "I don't think the comparison is ridiculous at all," he said. The timescales of past mass extinctions are "frighteningly similar to the timescales over which our current climate is changing. The cause might be different but the hallmarks are similar."[23]

So, let's define the plausible outer limits of our danger. What a large team of scientists in 2017 called a "biological annihilation" is already well under way, with half the planet's individual animals lost over the last decades and billions of local populations of animals already lost.[24] In 2018, researchers reported that some local populations of insects had declined 80 percent—and it's hard to wipe out insects. Even with more charismatic fauna, we don't notice the declines at first, because there are still plenty of pictures. (A study found that a French person sees more photos of lions in a year than there are actual lions left in West Africa.)[25] But these losses come from a multipronged assault: forests cleared for timber and farmland, coastal waters poisoned, tasty animals overhunted and overfished. And now we are, far more rapidly than ever before in Earth's history, filling the atmosphere with the precise mix of gases that triggered the five great mass extinctions. It's not that the planet can't eventually deal with this: over the very deepest time, all that carbon will eventually be turned into limestone in the ocean, and into oil and gas and coal, and eventually the cycle will repeat itself. If you back up far enough, nothing matters.

But perhaps we, of all creatures, shouldn't back up that far. Unlike the fishes of the Permian, we've been given a warning. Unlike the sauropods of the Cretaceous, we can do something about it. As Peter Brannen wrote in his history of the great cataclysms, "Thankfully we still have time"[26]—though clearly not much.

5

Privilege lies in obliviousness. (White privilege, for instance, involves being able to reliably forget that race matters.) One of the great privileges of living in the affluent parts of the modern world is that we've been able to forget that the natural world even exists. In our lifetimes, and the lifetimes of our parents, it's served mostly as a backdrop. A subdivision is named for what used to be there: Fox Ridge. A suburb is designed to hide the natural world: where, amid the curving streets, are the creeks? A great city seems to produce wealth out of thin air. This is illusion, of course, but powerful illusion. I didn't start to see through it until, as a young *New Yorker* reporter, I spent a year tracing every pipe and cable that entered and exited my Greenwich Village apartment, following the water mains and the electric lines and the sewers to their ultimate sources and destinations. In the process, I came to understand the remarkable *physicality* even of New York: the vast water tunnels built at unimaginable expense and danger and effort, the supply lines that stretched to the hydro dams of Hudson Bay and the oil wells of the Amazon Basin.

Given that it all works so smoothly, we can be forgiven for ignoring the natural world most of the time. It is safely underground or in the walls or out of sight, at the power plant or the waste treatment station. But that smooth operation, that humming efficiency, is beginning to buckle under the pressure of a changing climate. Hurricane Sandy came ashore in New York City, channeling the energy from a record-hot Atlantic Seaboard and riding the raised level of the sea—and suddenly FDR Drive was awash in whitecaps and the South Ferry subway entrance was a cascade of saltwater pouring onto the tracks below. Napa explodes in fire; Cape Town, parched by drought, rations drinking water.

Let's put aside, for the moment, the thought of mass extinction. Cataclysm on a geological scale is clearly possible; you can make an argument that the game is up. But even if that is our eventual due, life will first look and feel different. Life as we know it won't suddenly end, but it will be crimped; in many places, it already is. To use our metaphor, *the size of the board on which we're playing the game is going to get considerably smaller*, and this may be the single most remarkable fact of our time on earth.

That shrinkage is, in itself, novel. For all of human history we've been playing out the opposite story. We seem to have begun in Africa and then spread out, slowly at first and then much faster. For North Americans, the chief architects of the modern game, this expansion is close enough chronologically to be our national story. Many of us descend from Europeans who, fed up with the crowded conditions and religious strictures of the Old World, came to a new one. Upon arrival, they slaughtered or pushed aside the people already inhabiting this continent, and then imported boatloads of human chattel to do much of the work of building the "New World." Those basic and tragic facts haven't stopped us from deciding that the wealth created here was a sign of moral superiority: we Americans believe that we were particularly innovative and entrepreneurial and brave. In fact, however, our achievement was less the result of noble character, or even the constant willingness to oppress others, than it was a pure windfall. Those who settled North America vastly expanded the board on which Europeans were playing the game, and this new section was beyond compare.

As the great environmental historian Donald Worster points out, Columbus was looking for a new route to Asian wealth: silks, spices, and so on. What he found was so much better: "an unexpected abundance of space, land, soil, forests, minerals, and waters, an abundance that was almost free for the taking."[1] It was almost as if the Europeans had landed on a new planet—not one of the gaseous, hostile, barren planets of our solar system, but a planet like Europe or Asia, except mostly intact and undegraded. "Somewhere within its borders the United States offered almost everything that people wanted: the world's greatest expanse of prime soils; a supply of fresh water that seemed limitless (until one got to the western deserts); a forest cover that surpassed in quality, diversity, and utility that of any other nation; a vast renewable resource of furs and fish; and almost every mineral known to man," Worster observes.[2] Imagine, say, the impact of the invention of the internet on modern economic life, and then multiply it by many times. "The discovery of America, and that of a passage to the East Indies by the Cape of Good Hope, are the two greatest and most important events recorded in the history of mankind," wrote Adam Smith. They brought "dreadful misfortune" to the native inhabitants of those places, but by enlarging the game board, those new colonies raised "the mercantile system to a degree of splendour and glory which it could never otherwise have attained to."[3]

Eventually, of course, North Americans managed to fill up much of the new continent, but that didn't stop our expansion. By the 1890s, when Frederick Jackson Turner was declaring the frontier closed, another new continent was opening up, this one underground. Humans everywhere were quickly learning to burn fossil fuels, and so once again our range was expanding. Part of that expansion was literal: instead of being confined to the few villages where a horse or your feet could carry you, everyone was able to move about, a liberation from geography that changed everything, right down to whom you might marry. And cheap power led, at the turn of the century, to air-conditioning, which in turn meant that places once so hot as to be marginal were now "the Sun Belt."

But the biggest part of this new expansion was economic: everyone in the Western world now had access to, in essence, slaves who would do absurd amounts of manual work. A barrel of oil, currently about sixty dollars, provides energy equivalent to about twenty-three thousand hours of human labor. The great economist John Maynard Keynes once calculated that from "two thousand years before Christ down to the beginning of the eighteenth century, there was really no great change in the standard of living of the average man in the civilized centers of the earth. Ups and downs, certainly visitations of plague, famine and war, golden intervals, but no progressive violent change." What changed that was coal, and then oil and gas. All of a sudden, the standard of living was doubling every twenty or thirty years.

These were onetime gains. There are no new continents to be discovered, and even if enthusiasts chirp excitedly on about someday mining asteroids, that is a step down from discovering the vast forests of Appalachia. (Movie astronaut Matt Damon sort of managed to grow potatoes on Mars, but only because his own dung provided the necessary nutrients. That's not quite as good as Iowa topsoil.) We are, of course, discovering new kinds of energy. The solar panel, in particular (as we shall see in the final part of this book), is a variety of miracle, but a different kind of miracle from fossil fuel, which was so dense with power, so easy to transport. Our world has been broadening for centuries, and that broadening is, to a large degree, what we think of as normal and ordinary: if the economy doesn't grow larger each year, we now suffer as a result, because our systems, and our expectations, have become dependent on that growth. We play the game on a much larger board than our ancestors, and we play it with much more power.

But thanks to global warming, that broadening is now coming to an end, and a period of contraction is setting in. Instead of new continents to inhabit, our space is beginning to shrink. Our earth is large, but it is finite, and we're beginning to lose parts of it.

Sheer heat—heat alone, the most obvious effect of climate change—has begun to narrow the margins of our inhabitation. Nine of the ten deadliest heat waves in human history have happened since 2000.[4] Even places that define cool, like the Pacific Northwest, now see stretches where the heat soars into the triple digits, and 70 percent of the homes in Portland are now air-conditioned.[5] But in Portland, a hideous heat wave means that the city opens pet-friendly "cooling centers" stocked with board games. In India, by contrast, the average rise in temperature of a single degree Fahrenheit since 1960 has increased the chance of mass heat-related deaths by 150 percent.[6] Those heat waves are unbearably savage. In the summer of 2016, temperatures in cities in Pakistan and Iran peaked at slightly above 129 degrees Fahrenheit for a couple of days in July, the highest reliably recorded temperatures ever measured on planet Earth. (I just checked the oven in my kitchen, and you can set it at 130 degrees.) But as hot as those places were, it was a dry, desert heat. The same heat wave, nearer the shore of the Persian Gulf and the Gulf of Oman, combined triple-digit temperatures with soaring humidity levels to produce a heat index over 140 degrees Fahrenheit. In 2015, in Bandar-e Mahshahr, in Iran, the heat index reached 165 degrees, the highest ever witnessed on the planet.[7]

About a decade ago, Australian and American researchers set out to determine the maximum survivable combination of heat and humidity. They concluded that a "wet-bulb temperature" of 35 degrees Celsius set the limit—that is, when temperatures passed 35 degrees Celsius (95 degrees Fahrenheit) and the humidity was above 90 percent, "the body can't cool itself and humans can only survive for a few hours, the exact length of time being determined by individual physiology." That's because evaporation off the skin slows down in the humidity; you can't cool yourself by sweating. "Not even the fittest of humans can survive, even in well-ventilated shaded conditions, when the wet bulb temperature stays above 35," said one of the scientists. They went on to conclude that about 1.5 billion people, a fifth of humanity, lived in a crescent-shaped area at high risk of such temperatures as the planet warmed.

That includes some of the world's most densely populated regions, in India, Pakistan, and Bangladesh, as well as those Middle Eastern cities along the sea. In these places, extreme heat waves that now happen once every twenty-five years will become "annual events with temperatures close to the threshold for several weeks each year, which could lead to famine and mass migration."[8] Because, of course, these are precisely the places where most of the population works outdoors. In 2018, new research made it clear that the North China Plain, with 400 million residents, fell squarely in this red zone. "This is going to be the hottest spot for deadly heat waves in the future," one MIT professor explained. "Continuation of current global emissions may limit the habitability of the most populous region of the most populous country on earth."[9]

So: the world we've known is quickly being replaced by a new one, and this planet is effectively closer to the sun. As a result, by the 2070s, tropical regions that now get one day of truly oppressive heat a year can expect 100 to 250 days. By 2100, the most recent study notes, "even under the most optimistic predictions for emissions reductions, experts say almost half the world's population will be exposed to potentially deadly heat for twenty days a year."[10] "Lots of people would crumble well before you reach" these maximum readings, one of the analysts explained. "They'd run into terrible problems." The result, he added, would be "transformative for all areas of human endeavor—economy, agriculture, military, recreation."[11] Already, increased heat and humidity have cut the amount of work people can do outdoors by 10 percent, and that effect should double by midcentury.[12] A new report on Florida farmworkers found "more and more people that have dehydration" as a result of rising temperatures. Undocumented migrants are "especially vulnerable, as they are less likely to demand rest, shade or water for fear of retaliation."[13] In many places, it will simply be too muggy for humans to do the work of humans.

The summer of 2018 was the hottest ever measured across large stretches of this planet. Africa recorded its highest temperature ever in

June, the Korean Peninsula in July, and Europe in August; in America, Death Valley produced the hottest month ever seen on our continent. The world saw the warmest night in history, when the mercury in one Omani city stayed above 109 degrees Fahrenheit till morning. In Algeria, a *New York Times* reporter found employees at a petroleum plant simply walking off the job as the temperature neared 124 degrees. "We couldn't keep up," said one worker. "It was impossible to do the work. It was hell." In Nawabshah, Pakistan, the heat set a new local record, 122 degrees Fahrenheit, and "shops didn't bother to open. Taxi drivers kept off the street to avoid the blazing sun."[14] In Montreal, where a heat wave had killed seventy-seven people, a homeless man described his life: he moved two or three blocks at a time, from one air-conditioned mall to the next, waiting to be turned out. "We need more water fountains in the park," he told reporters for the *Guardian*, who also interviewed a student in Cairo, where the temperature was a mere 104 degrees. His extended family had saved up to buy one air-conditioning unit for the living room, and "now that's where everyone spends their day—preparing food, watching TV, playing or studying."[15] In other words, their world had shriveled to a single room. When a city gets that hot, as one reporter put it, "the pavements are empty, the parks quiet, entire neighborhoods appear uninhabited. Nobody with a choice ventures outside."[16]

And as with the heat, so with the oceans. Their rise is driving people away from the places we've always inhabited. The same Asian peasant farmers having to cope with hideous heat in the fields are also watching as saltwater wrecks those soils—tens of thousands now evacuate Vietnam's sublimely fertile Mekong Delta annually. You don't have to search to find the scary details, given that most coastal communities have at least begun to study the possible impacts. In one week at the end of 2017, without making any special effort, I came across stories from Louisiana, where government officials were already finalizing a plan to move thousands of people from rising seas ("Not everybody is going to be able to live where they are now and continue their way of life," said one state official);[17] from Hawaii, where a new study was predicting that, over the next few decades, thirty-eight miles of coastal roads would be chronically flooded and impassable, "jeopardizing critical access to many communities";[18] from Jakarta, Indonesia's mega-city, where a rising Java Sea earlier that month had briefly turned "streets into rivers and brought this vast area of nearly 30 million residents to a virtual halt";[19] and from Boston, where a simple nor'easter in the first days of 2018 managed to flood some of the city's priciest neighborhoods, floating Dumpsters and sedans through the Financial District. "If anyone wants to question global warming, just see where the flood zones are," Boston's mayor said. "Some of those zones did not flood thirty years ago."[20]

If you trace out the area that's ten meters above current sea levels, it covers only 2 percent of the earth's land area, so the game board won't shrink *enormously* from sea level rise. But that 2 percent of the surface contains 10 percent of the people, and generates 10 percent of the gross world product.[21] And it's not defensible, not most of it—no one is going to pay to build a seawall around the Bengali coast; or to defend Accra, the capital of Ghana, which already floods during storms. "On the outskirts of Lomé, the capital of Togo, rows of destroyed buildings line the beaches," Jeff Goodell reports.[22] Anyone want to estimate how much money the world is likely to spend defending the capital of Togo? "Like it or not, we will retreat from most of the world's non-urban shorelines in the not very distant future," the Duke University sea level rise expert Orrin Pilkey wrote in 2016. "Our retreat options can be characterized as either difficult or catastrophic. We can plan now and retreat in a strategic and calculated fashion, or we can worry about it later and retreat in tactical disarray in response to devastating storms. In other words, we can walk away methodically, or we can flee in panic."[23]

As some people flee more water (in the form of humidity or of sea level rise), others will be moving because there's too little. Remember: wet areas get wetter as the planet warms, but arid areas get even more droughty. In late 2017 a study estimated that by 2050, even if the world manages to hit the Paris climate target of "only" a two-degree Celsius rise in temperature, a quarter of the earth would experience serious drought and desertification. "Our research predicts that aridification would emerge over about 20 to 30 percent of the world land surface," said the study's lead author. Another study from the same year found that as hotter days led to more evaporation, corn and soybean yields across the U.S. Grain Belt could fall by 22 to 49 percent. Extensive irrigation could help—except that we've already overpumped the aquifers that lie beneath most of the world's breadbaskets.[24] Some Americans can still remember what drought-driven dislocation looks like: Okies piled into rattling pickups streaming out of the Dust Bowl and into California's pastures of plenty (and one Harvard researcher recently predicted that America's climate migration will be twice the size of that Depression-era exodus).[25] But now, as we've seen, even reliable escape routes are blocked. California's snowpack keeps dwindling as hot, dry years pile up; the state faces a drop of as much as 70 or 80 percent in its water supply.[26]

Even in those places where you'd expect the field of play to be expanding, we're seeing the opposite. Warmer temperatures should make the Arctic into the new Kansas, right? Here's how Rex Tillerson cheerfully put it, back when he was the CEO of Exxon: "Changes to weather patterns that move crop production areas around—we'll adapt to that." Except Iowa is Iowa not just because of the temperature. There's also that topsoil, none of which you find once you move north; instead, the ground there is underlain with ice. And as that permafrost melts, it spews more carbon into the atmosphere—no small matter, given that permafrost makes up one-fifth of the Northern Hemisphere. But that thawing layer also cracks roads, tilts houses, and even uproots trees to create what scientists call "drunken forests." Economic losses from a warming Arctic could approach $90 trillion over the course of the century, far outweighing the gains from easier shipping lanes, according to ninety scientists who released a joint report in 2017.[27]

You get a sense of why by looking at particular places: Churchill, Manitoba, say, on the edge of Canada's Hudson Bay. A single rail line connects it to the lower world, but in the spring of 2017, record floods washed away much of the track. The company that owns the rail line says it can't justify the price of fixing it, "particularly in a warming climate." To cancel its contract, the company declared what lawyers call a "force majeure," an unforeseen event beyond its responsibility. "To fix things in the era of climate change, well, it's fixed but you don't count on it being the fix forever," an engineer for the company explained. "Things are changing that we can't arrest or change or govern." Even construction of a new research center to study the effects of climate change ceased when the train shut down.[28]

If you have enough money you can ward off anything for a while. The Canadian government reopened the rail line in the summer of 2018 at the cost of $117 million—about $130,000 per resident of Churchill. But next time? Churchill "claims a mythic place in the Canadian psyche," up there at the end of the rail line. And so do many of the other places that we may abandon before too long. Fort Sumter? The Kennedy Space Center? Mar-a-Lago? It's worth noting that those Iraqi cities with the increasingly impossible temperatures sit close to where biblical scholars place the Garden of Eden. In 2018, Scottish archaeologists reported that thousands of prehistoric sites—stone circles, Norse halls, Neolithic tombs—were at risk from rising seas. Each tide washes away artifacts—washes away our history.[29]

Lots of people already hesitate to walk across a grassy meadow because hot weather has spread ticks bearing Lyme disease. On plenty of beaches, people now sit stranded on the sand because jellyfish, which thrive as warming seas kill off other marine life, have taken over the waves. The planet's diameter will remain eight thousand miles, and its surface will still cover two hundred million square miles, but the earth, for humans, has begun to shrink, under our feet and in our minds.

6

Climate change has been a public issue for thirty years. It's true that there were scientific reports and presidential memos scattered across the preceding decades that warned we might face trouble, and the pace accelerated in the 1980s. This early phase of climate politics was nicely outlined by Nathaniel Rich in a recent special issue of the *New York Times Magazine.* But the important thing to remember is that it all happened behind closed doors, in meetings confined to a few scientists and officials.[1] The world, its leaders and its citizens, effectively knew nothing of the threat until the hot June day in 1988 when a mid-career NASA scientist named James Hansen testified before a Senate committee that "the greenhouse effect has been detected and it is changing our climate now."[2]

In the weeks that followed, members of Congress introduced the National Energy Policy Act to "address . . . heat-trapping gases produced in burning fossil fuels." The world's atmospheric scientists announced the formation of the Intergovernmental Panel on Climate Change to track the crisis. And Vice President George H. W. Bush, in the midst of a successful campaign for the White House, announced that he would "fight the greenhouse effect with the White House effect." It looked like America meant business, that a response was starting to take shape.

But, as it turned out, that didn't really happen. In the three decades since, global carbon emissions have nearly doubled. More than half of all the greenhouse gases emitted since the start of the Industrial Revolution have spewed from exhaust pipes and smokestacks since 1988.[3] In all but one year since 1988, we've burned more fossil fuel than the year before, and the exception was 2009, when the economy fell off a cliff.

Which is to say, Donald Trump is a horrible human being who has done all that he can think of to retard progress on climate change, but it's not his fault the planet is overheating.

The tepid response to what scientists were quickly calling the greatest challenge humans have ever faced was in certain ways predictable. I remember, in 1988, when I was finishing *The End of Nature*, interviewing a political scientist who described it as "the problem from hell." There were too many different interests, he said, from too many parts of the world. Fossil fuel was at the center of the world's economy, involved in every moment of a modern day—and yet it was the very thing that was killing us. It was as if a doctor had told you that your chief problem was that your heart and lungs were pumping poison through your body. There was nothing we could do, he said, at least not in the time we had.

His assessment has so far turned out to be correct, and it's worth taking a look at a few of the reasons for this.

One is simple inertia, never to be underestimated. Anthropologists talk regularly about how we evolved to deal with a snarling tiger emerging from behind a tree: we're geared to short-term thinking because that's what allowed us to survive; whereas tomorrow was always a problem for tomorrow. But think how amplified that is when you have a literal investment in the present—when, say, a few hundred billion dollars are at stake. The mayor of Miami Beach, whose streets already flood regularly, told a crowd at the city's one-hundredth-anniversary gala that "I believe in human innovation. If thirty or forty years ago, I'd told you that you were going to be able to communicate with your friends around the world by looking at your watch or with an iPhone, you'd think I was out of my mind."[4]

So: the snarling tiger is not actually eating us at this precise moment. Anyway, in thirty years maybe there will be an app to deal with snarling tigers. It's easy enough to make fun of this kind of reaction, but in fact, it's how almost all of us have reacted. We've gone on more or less as usual. We literally don't want to hear about it.

Even those politicians who have wanted to do something about it have wanted to do something easy. They've looked, naturally enough, for relatively small steps they could take that would provoke as little outcry as possible. And they've argued, with the persuasiveness of people who must get elected, that that's all they can do. "Part of my job is to figure out what's my fastest way to get from point A to point B—what's the best way for us to get to a point where we've got a clean energy economy," Barack Obama explained toward the end of his last year in office. "And somebody who is not involved in politics may say, 'Well, the shortest line between two points is just a straight line; let's just go straight to it.' Well, unfortunately, in a democracy, I may have to zig and zag occasionally, and take into account very real concerns and interests."[5]

Obama's "zigs and zags" are illustrative—they neither saved the day nor wrecked our chances for survival, but they did reflect just how hard even a good-faith effort on this "problem from hell" can be. Environmentalism was not his main concern, but Obama understood that climate change was important: the night he clinched the Democratic nomination in 2008, he said that "this was the moment when the rise of the oceans began to slow and our planet began to heal." And it seemed like his timing was, as usual, impeccable. His term in office coincided with the large-scale advent of fracking. Suddenly, it appeared that America had an enormous supply of natural gas that could be liberated easily from shale in Texas and the Appalachians. For environmentalists, that initially seemed like very good news, because when you burn natural gas it gives off half as much carbon dioxide as the coal that powered most of the nation's, and the world's, electric supply. For

Obama, it was a godsend. He could reduce America's carbon emissions with minimal upset. The big oil companies controlled much of the natural gas supply, and they wouldn't protest. Utilities liked it because it kept their infrastructure essentially intact—in many cases, you could actually just convert the old coal-fired power plant to burn the new supplies of gas. And though the use of natural gas did continue the slow-motion implosion of the coal industry, that was more than offset by the way gas helped spark the dead-in-the-water economy Obama had inherited. Manufacturing jobs were returning from overseas, attracted by the newly abundant energy. In his 2012 State of the Union address, the president declared that new natural gas supplies would not only last the nation a century, but also create six hundred thousand new jobs by decade's end. He boasted that under his administration, they'd "added enough new gas pipelines to encircle the Earth and then some."[6]

So, no one wanted to hear the chemists when they began to raise an uncomfortable issue about natural gas. It's indeed true, they said, that methane produces only half as much carbon as coal when you burn it. But if you *don't* burn it—if it escapes into the air before it can be captured in a pipeline, or anywhere else along its route to a power plant or your stove—then it traps heat in the atmosphere about eighty times more efficiently than carbon dioxide. Two Cornell professors, Robert Howarth and Tony Ingraffea, produced a series of elegant papers showing that if even a small percentage of fracked gas leaked, maybe as little as 3 percent, then it would do *more* climate damage than coal. And their preliminary data showed that leak rates could be at least that high—that between the fracking operations and the thousands of miles of pipes and the compressor stations, somewhere between 3.6 and 7.9 percent of methane gas from shale-drilling operations actually escapes into the atmosphere. In June 2018 a new study found that the amount of methane leaking from oil and gas fields was 60 percent higher than the official EPA estimate.[7] Eventually, in fact, satellite data showed that U.S. methane emissions had spiked by 30 percent since 2002. That meant that *total* U.S. greenhouse gas emissions (carbon dioxide and methane combined) had barely budged during the Obama years. In fact, they might have gone up. Carbon dioxide emissions declined, yes, but they were offset by the rising spew of methane. In other words, a crucial decade had been wasted—worse than wasted, because all the new drill rigs and pipelines and gas-fired power plants will be in operation for decades to come.

It's not, at some level, Obama's fault. He was elected to run a political and economic system based on endless growth. He feared that if he upset it too much he wouldn't be reelected, which would have done no one any good. (Though it's worth noting that he continued to boast about it even after leaving office. In November 2018 he told a Texas audience that during his term America had passed Russia and Saudi Arabia as the biggest oil and gas producer. "That was me, people," he said.)[8] It's the same around the planet, though different leaders fear different things: angry central committees, upset oligarchs, mobs angry at higher gasoline prices. Against this kind of institutional inertia, even charisma counts for little. Consider the handsomest, most progressive, most apparently "woke" leader on the planet, Canada's Justin Trudeau. He's far more outspoken on climate change than Obama ever was: it was Canadian diplomats, at his insistence, who persuaded the nations of the world to lower their preferred climate target from 2 degrees Celsius to 1.5 degrees in the final days of the Paris negotiations. "There is no country on the planet that can walk away from the challenge and reality of climate change," Trudeau told the UN General Assembly. "We have a responsibility to future generations and we will uphold it."[9] And yet, Trudeau's country contains one of the two largest deposits of tar sands on earth, that vast swath of Northern Alberta that can, at great cost to water and forest, be mined for sludgy oil. And Trudeau refuses to slow its expansion. When a pipeline company tried to back out of building a new pipeline to the British Columbia coast in 2018, Trudeau *nationalized* it, committing more than ten billion dollars in taxpayer money.

Like so many politicians, he turned out to be unwilling to relinquish the power that oil represents. In the spring of 2017, Trudeau told a cheering group of Houston oilmen that "no country would find 173 billion barrels of oil in the ground and just leave them there." And yet, just leaving them there is exactly what he'd have to do if he were even slightly serious about taming climate change. If we burn that 173 billion barrels of oil, the carbon dioxide will take us 30 percent of the way to the 1.5 degree target that Trudeau had insisted on in Paris. That is, one nation with one-half of 1 percent of the planet's population is laying claim to a third of the atmospheric space between us and disaster.

If a leader such as Justin Trudeau can't manage more bravery, then what hope is there? Especially given that time is so short. Unlike the other issues that politicians deal with, this one *can't* proceed via the slow zigs and zags that Obama described. Climate change is not a normal political negotiation between different interests, where compromise makes obvious sense. Climate change is a negotiation between human beings and *physics*, and physics doesn't compromise. Past a certain point, there's no more room for maneuver.

That point is clearly upon us: it's not a good sign that the largest physical structures on our planet, its ice caps and barrier reefs and rain forests, are disappearing before our eyes.

So: problem from hell. Governments prefer to evade it. Human psychology is not designed to cope with it. It's happening too fast.

And yet, one can't help but think, *We've faced problems from hell before.*

In the twentieth century we faced Hitler. Most Americans initially wanted to deny the threat he represented—the U.S. Chamber of Commerce argued against lending ships to the British to fight him. The eventual effort cost more money than anything anyone had ever done; millions and millions of people had to upend their lives; four hundred thousand Americans had to die (and ten million Soviet soldiers)—but we did it. So why haven't we taken on the similar-size crisis of our time?

True, that generation had Pearl Harbor to push them into the fight against the Axis. But we've had Katrina and Sandy and Harvey. And it's not as if we've had no alternatives—by this point, as we'll eventually discuss, sun and wind are the cheapest way to generate power on planet Earth. No one has to die in this fight. As a task, it would be a far easier one than fighting a world-spanning war.

So, yes, climate change is a very hard problem. But there's something more going on here than the usual inertia.

7

There should be a word for when you commit treason against an entire planet.

In July 1977, one of Exxon's senior scientists, a man named James F. Black, addressed many of the company's top leaders. Speaking at the oil giant's New York City headquarters, he flipped through slide after slide showing some of the earliest research under way on what was then called the greenhouse effect. He concluded: "There is general scientific agreement that the most likely manner in which mankind is influencing the global climate is through carbon dioxide release from the burning of fossil fuels."[1] A year later, he spoke to a larger pool of the company's executives. Independent researchers, he said, estimated that a doubling of the carbon dioxide concentration in the atmosphere would increase average global temperatures by 2 to 3 degrees Celsius (3.6 to 5.4 degrees Fahrenheit) and as much as 10 degrees Celsius (18 degrees Fahrenheit). Rainfall might get heavier in some regions, and other places might turn to desert.[2]

That is to say, ten years before James Hansen's Senate testimony made climate change a public issue, Exxon, the world's largest oil company and, indeed, in those days, the world's largest company period, understood that its product was going to wreck the planet. We know this because of extraordinary reporting, first from a Pulitzer Prize–winning website called InsideClimate News, and then from the *Los Angeles Times* and the Columbia Journalism School. What they discovered, through deep dives into company archives and interviews with former employees, is the most consequential cover-up in human history.

Fossil fuel corporations had worried at least a little about climate change for a long time. As early as 1959, at a symposium called "Energy and Man," organized by the American Petroleum Institute to mark the centenary of the global oil business, the physicist Edward Teller told the industry's most important executives, "Carbon dioxide has a strange property. It transmits visible light but it absorbs the infrared radiation which is emitted from the earth." The temperature, Teller predicted, would rise, and when it did, "there is a possibility that the icecaps will start melting and the level of the oceans will begin to rise."[3] But these kinds of warnings were easily ignored. Rachel Carson had yet to begin knocking the shine off modernity; her book *Silent Spring* would come in 1962. More important, global warming was mere speculation, because nobody had the computing power to model something as complicated as the climate. It's true that as early as 1968, the president's science advisor warned the annual meeting of the nation's utilities that changes in carbon dioxide "might produce major consequences on the climate—possibly even triggering catastrophic effects such as have occurred from time to time in the past."[4] But it's also true that nobody knew for sure.

By the late 1970s, though, when Exxon took on the issue, the greenhouse effect had moved from vague possibility to something much more ominous: "Present thinking holds that man has a time window of five to ten years before the need for hard decisions regarding changes in energy strategies might become critical," James Black told the assembled Exxon executives. Two years later, the company's scientists, in a document given wide distribution among senior executives, said "there is no doubt that increases in fossil fuel usage and decreases of forest cover are aggravating the potential problem of increased CO_2 in the atmosphere." The American Petroleum Institute assembled an industry task force, with representatives from Exxon, Texaco, Shell, Gulf, and others, to "look at emerging science, the implications of it, and where improvements could be made, if possible, to reduce emissions."[5] Exxon decided to spend millions of dollars on the research effort—after all, its product was carbon, and it needed to understand it. Among other things, it outfitted an oil tanker, the *Esso Atlantic*, with carbon dioxide detectors in an effort to measure how fast the oceans could absorb excess carbon, and it hired mathematicians to build more sophisticated climate models. By 1982, it had concluded that even the company's earlier dire estimates were probably too low. That year, in a corporate document marked "not to be distributed externally" but given "wide circulation to Exxon management," the company's scientists concluded that heading off global warming would "require major reductions in fossil fuel combustion." Otherwise, it concluded, "there are some potentially catastrophic events that must be considered." Delay, it warned, was dangerous. "Once the effects are measurable, they might not be reversible."[6]

We know that Exxon executives took these warnings seriously. Internal documents show that the company (and other oil giants) built their new oil drilling platforms with higher decks to compensate for the sea level rise they now knew was coming. In the Arctic, a team assigned to investigate the effects of warming concluded that "global warming can only help lower exploration and development costs" in the Beaufort Sea. As the team's leader told an industry conference in 1991, "greenhouse gases are rising due to the burning of fossil fuels. . . . No one disputes this fact." As a result, the team predicted, drilling season in the Arctic would lengthen from two months to as much as five months, which is in fact what has happened.[7]

It wasn't just Exxon that knew. In the autumn of 2018 new documents emerged showing that Shell scientists had predicted in the late 1980s that carbon dioxide levels could double as early as 2030, and predicting an increase in "runoff, destructive floods, and inundation of low-lying farmland." All in all, Shell's experts said, "the changes may be the greatest in recorded history."[8] Take a minute to think through the implications of these exposés. By 1988, when James Hansen made global warming a public issue, the oil companies knew that he and the other researchers were right. They were, in fact, using Hansen's NASA climate models to figure out how low their drilling costs in the Arctic would eventually fall. So, imagine what would have happened if they had merely told the truth. Imagine, in July 1988, after Hansen told the U.S. Senate that global warming was very real and very dangerous, that the CEO of Exxon had simply said, "Our research bears this out. It appears to be true." That seems the least that any system of morality would demand. And it would not have been necessarily economically destructive. Indeed, with their advance knowledge, companies such as Exxon would have had the early inside track on building the energy economy of the future. As early as 1978, one manager at Exxon had said, "This may be the kind of opportunity that we are looking for to have Exxon technology, management and leadership resources put into the context of a project aimed at benefitting mankind."[9]

Had Exxon and its peers taken that course, history, *geological* history, would have been very different. No one would have said, "Oh, Exxon is just being alarmist." Everyone would have acknowledged the depth of the trouble, and gotten to work. It wouldn't have made the work easy—all the obstacles I described in the last chapter, from inertia to human psychology, would still have existed. Thirty years later, climate change would not have been *solved*. But as with the hole in the ozone layer, we would have taken large strides. We'd be on the way to a solution; the crisis would be abating.

But that's not what happened, of course, because this wasn't the hole in the ozone. In that case, the culprit was a small class of gases for which the manufacturers had available substitutes. Now the culprit was fossil fuel, the most lucrative substance on earth. And so, a month after Hansen's testimony, Exxon's public affairs manager recommended in an internal memo that the company "emphasize the uncertainty" in the scientific data about climate change.[10] Thus began the most consequential lie in human history. Within a year, Exxon, Chevron, Shell, Amoco, and others had joined together to form what they called the Global Climate Coalition, "to coordinate business participation in the international policy debate" on climate change. The GCC hired veterans of earlier fights against the tobacco industry; it even hired the company that had spearheaded the attack in the 1960s against Rachel Carson. It coordinated with the National Coal Association and the American Petroleum Institute on a "grassroots letter and telephone campaign to prevent a proposed tax on fossil fuels" and produced a video insisting that more carbon dioxide would "end world hunger." It also ginned up opposition to the 1997 Kyoto Protocol, the first global effort to do anything about climate change.

Two months before the Kyoto meeting, Lee Raymond (Exxon's president and CEO, and the man who had had oversight responsibility for the science department that in the 1980s produced the unambiguous findings about climate change) gave a speech in Beijing at the World Petroleum Congress that belongs on the short list of the most irresponsible speeches any American has ever delivered. He insisted that the Earth was cooling, said that the idea that cutting fossil fuel emissions could have an effect on the climate "defied common sense," and declared that, in any event, it was "highly unlikely that the temperature in the middle of the next century will be affected whether policies are enacted now, or twenty years from now." Remember, Exxon's own scientists had shown each of these premises to be wrong; Exxon itself was basing its own corporate decision making on that science. It knew the Beaufort Sea was melting; it was building its drillings rigs high enough to take on a rising sea. It just wasn't telling the rest of us.

Environmental pressure eventually forced the Global Climate Coalition to disband. BP and Shell left after European green groups mounted savage campaigns, and many of the American companies eventually dropped out. But it was a pyrrhic victory: by then, the damage had been done. I remember lurking on the edge of the Kyoto convention center the morning after a long night of negotiation had finally produced a tentative accord. Imperfect and limited though that accord was, it seemed to me that the momentum had swung in the direction of actually fighting climate change. But I was standing next to a lobbyist who had been coordinating much of the fight against the accord, and as delegates cheered and clapped, he turned to me and said, "I can't wait to get back to Washington, where we've got this under control." Many of the Global Climate Coalition hacks went to work inside the George W. Bush administration. Nine days after Bush was inaugurated, Lee Raymond came for a visit with his old friend Vice President Dick Cheney, who had just stepped down as the CEO of oil-drilling giant Halliburton. Raymond apparently helped persuade Bush to abandon his campaign promise to treat carbon dioxide as a pollutant, and within the year, Bush's pollster, Frank Luntz, had produced an internal memo that canonized the strategy the GCC had hit on a decade earlier: "Voters believe that there is no consensus about global warming within the scientific community," Luntz wrote. "Should the public come to believe that the scientific issues are settled, their views about global warming will change accordingly. Therefore, you need to continue to make the lack of scientific certainty a primary issue in the debate."[11]

The strategy worked, exactly as well as they needed it to. As late as 2017, pollsters found that almost 90 percent of Americans didn't know there was a scientific consensus on global warming.[12] Lee Raymond stepped down in 2006, taking a retirement package worth $400 million, after the company posted the greatest corporate profits in history. His successor, Rex Tillerson, was slightly less confrontational, and willing at least to grant that climate change might be real, though, at shareholder meetings, he continued to downplay the threat. ("What if it turns out our models are lousy and we don't get the effects we predict?" he asked in 2015.)[13] And the company continued to fund climate change deniers and front groups. One, the Competitive Enterprise Institute, put out a TV commercial titled "Carbon Dioxide: They Call It Pollution, We Call It Life." Another, the Heartland Institute, which Exxon had helped found back in the 1990s, erected billboards comparing climate scientists to famous serial killers such as the Unabomber and Charles Manson. Exxon also signed a $500 billion deal to explore for oil in the Russian Arctic (exploration that was possible only because the area was rapidly melting), and for his billions, Tillerson was officially awarded the Russian Order of Friendship in a ceremony at Vladimir Putin's villa. No matter the danger posed by fossil fuel, Exxon was never going to let anything change. As Tillerson told his last shareholder meeting, the planet "is going to have to continue using fossil fuels, whether they like it or not."[14]

Tillerson eventually went to work, of course, as the U.S. secretary of state, and his boss, Donald Trump, was the perfect example of just how well the company's PR strategy had succeeded. Trump, his news diet centered on the Fox News cable network that the climate deniers had so assiduously cultivated, believed that global warming was a "hoax invented by the Chinese" to cripple American manufacturing. (He further believed that polar ice caps "are at a record level.")[15] As a result, he pulled America out of the Paris climate accords, meaning that the nation that had poured the most carbon into the planet's atmosphere was now the only country not willing even to pretend to take action to stop the crisis. Meanwhile, at Exxon, business continued as usual. Tillerson's successor, Darren Woods, fended off shareholder pressure by agreeing to have his executives write a report disclosing the company's "climate risk." When that report was released in the winter of 2018, it found that Exxon faced no need to change at all. The team at InsideClimate News, which had broken the original story about Exxon's lies, summarized the company's statement: "Exxon insists it would be able to produce all the oil in its existing fields and to keep investing in new reserves."[16]

It's by no means clear that this three-decade campaign of deception and obfuscation is illegal. Exxon has always insisted that it has "tracked the scientific consensus on climate change, and its research on the issue has been published in publicly available peer-reviewed journals."[17] In any event the First Amendment preserves one's right to lie, though in the fall of 2018 the New York attorney general, Barbara Underwood, filed suit against Exxon for lying to investors, which *is* a crime. In January 2019, the Supreme Court ruled the company had to turn over millions of pages of internal documents to the Massachusetts attorney general, Maura Healey, so we will learn more.

What is certain is that this disinformation campaign cost us the human generation that might have made the crucial difference in the climate fight. Alex Steffen, an environmental writer, coined the term *predatory delay*, "the blocking or slowing of needed change, in order to make money off unsustainable, unjust systems in the meantime." Climate change, and the behavior of the oil companies, is the prime example. "Had we begun cutting global emissions in 1990, we could still have tackled the climate crisis with confidence," he writes. "The back of the envelope take is that we could then have cut emissions by something on the order of one quarter per decade and kept within our CO_2 budget." It "wouldn't have been child's play," but "well-understood incremental regulatory reforms and well-designed carbon trading or pricing systems" would have done the trick. But now, rushing up on 2020, and after three decades of soaring carbon emissions, meeting even the modest targets of, say, the Paris climate accord has become nearly impossible. "The path now is steep as hell—the new curve we're on demands downright disruptive emissions cuts" of as much as 50 percent a decade. As the geophysicist Michael Mann put it, "what would have been a bunny slope was now a double black diamond." That means, Steffen explained, that "climate action can no longer be orderly, gradual, or even continuous with our expectations."[18] By early 2018, the view from the analysts calculating transition efforts around the world was stark. "It's not fast enough," said one. "It's not big enough. There's not enough action."[19]

I always knew that the climate fight would be hard, harder than anything humans had ever done before. In *The End of Nature*, I said I doubted we'd make progress fast enough to hold off the wholesale transformation of our planet. But I was twenty-eight then, and unable to conceive that even as I was writing the book, a few of the most powerful people on Earth were sitting down to hatch a lie that would make the task infinitely harder. I've lived the last thirty years inside that lie, engaged in an endless debate over whether global warming was "real"—*a debate in which both sides knew the answer from the beginning.*

It's just that one of those sides was willing to lie. And so, we need to understand where that lie came from.

PART TWO

Leverage

8

I said I was a naïve young man. Let me prove it again.

A year out of college, reporting for *The New Yorker*, I went to a little Mississippi Delta town called Tunica. I was there to write about a stretch of shacks along a small and fetid creek called Sugar Ditch. The town's residents, all black, lived without plumbing or running water. Jesse Jackson had called it "America's Ethiopia," and soon it became a scandal. Eventually, *60 Minutes* arrived to do a special report. The neighborhood stank in the hot sun, the walls of the shanties were alive with bugs, and the people who lived there seemed as beaten down as it was possible for me to imagine. A stone's throw—a stone's underhanded lob—away were the suburban streets of the town's white residents, which looked like the cul-de-sacs and split-level ranches I'd grown up among.

I wasn't naïve about the existence of poverty—in those days, I was also running a homeless shelter in the basement of my Manhattan church. I was naïve about the *future*. I thought we'd do something about homelessness, and I knew for sure that Tunica was an aberration, something out of the distant past. That's how everyone else covered the story, too: somehow this overlooked remnant of a sadder day had survived, a sort of anti-Williamsburg depicting the worst of the sharecropping era. It was an echo, as much an oddity as an embarrassment. And the next year, right on schedule, the federal government began razing the homes and replacing them with newly built apartments. After all, as the U.S. Department of Housing and Urban Development pointed out at the time, more than fifty diseases were associated with contact with human waste. Diseases from human waste obviously made no sense in a rich country.

My naïveté stemmed naturally from the fact that I grew up at precisely the moment when America was making huge strides toward reducing inequality, when it seemed that the obvious task was to make our world fairer. I was born in 1960, between the New Deal and the Great Society. My childhood featured the civil rights movement and the women's movement. I thought that was what politics was about. The year I graduated from high school, 1978, was the year the top 1 percent of Americans saw their share of the nation's wealth fall to 23 percent.

Which, as it turned out, was as low as it would ever go. Since then, the wealthy's share of the take has doubled. CEOs made less than 20 times as much as the average worker when I was born; now they make 295 times as much.[1] And with the rapid rise of inequality has come a rebound of truly gross poverty, the kind I'd seen in Tunica and imagined was a relic. In 2017 the United Nations sent its special rapporteur on extreme poverty and human rights to tour America. After a two-week visit, this Australian expert concluded that "for one of the world's wealthiest countries to have forty million people living in poverty, and over five million living in 'Third World' conditions is cruel and inhuman."[2] He recounted many horrors—fourteen thousand homeless people in San Francisco arrested for public urination when the ratio of toilets to people in the city's Skid Row "would not even meet the minimum standards the UN sets for Syrian refugee camps"; and the people he'd seen without any teeth because adult dental care isn't covered by Medicaid—but he focused most closely on the prevalence of hookworm in rural America. Hookworm (*Necator americanus*) is a parasite that enters the body through the soles of bare feet, and eventually attaches itself to the small intestine, "where it proceeds to suck the blood of its host. Over months or years it causes iron deficiency and anemia, weight loss, tiredness and impaired mental function, especially in children, helping to trap them into the poverty in which the disease flourishes."[3]

In one county that researchers from the Baylor College of Medicine studied (Lowndes County, a hotbed of the Alabama civil rights fight in the 1960s), *34 percent* of residents tested positive for traces of the disease. As in Tunica decades before, raw sewage was at issue—reporters following up on the study found trailer parks where sanitation was simply a cracked pipe leading into the woods, crisscrossing the pipe that carried water back into a trailer. "The open sewer was festooned with mosquitoes, and a long cordon of ants could be seen trailing along the waste pipe from the house," the *Guardian* reported. "At the end of the pool nearest the house the treacly fluid was glistening in the dappled sunlight—a closer look revealed that it was actually moving, its human effluence heaving and churning with thousands of worms." Researchers estimated that twelve million Americans might now be suffering from neglected tropical diseases such as hookworm across warm parts of the country.[4]

This is not a book about poverty and inequality. There are many such books, and there need to be many more, because real people are having their real lives blighted every day. But poverty and inequality and injustice are not going to end the human game. The pendulum can, and will, eventually swing back the other way, toward the kind of efforts at egalitarianism that marked my childhood. It will happen through reform or through revolution; that's a repeating rhythm in human affairs. We need to speed up that rhythm, and done right, the fight against inequality meshes powerfully with the fight against more existential threats such as climate change. That's why, at 350.org, we talk a great deal about "climate justice," convinced that it's both right and smart to work most closely with communities on the front lines of environmental damage. It's why we're excited by efforts such as the Poor People's Campaign, or the Leap Manifesto that Naomi Klein produced with an assortment of labor unions and indigenous people. It's all the same struggle.

But for this book, the rapid rise in poverty and inequality will primarily serve as a marker of who we are right now and how we got here; what we care about; how we understand the world. And for those purposes, the most important part of this "we" is the people in power, either formal or informal, who have allowed this poverty and inequality to shoot up over the last four decades. Not allowed—*encouraged*. Within days of the UN special rapporteur's report on extreme American poverty, the U.S. Congress responded by passing a massive tax cut that virtually every economist predicted would make that inequality much worse. As the UN expert noted in his official report to the world body, "The strategy seems to be tailor-made to maximize inequality. . . . It seems driven by contempt, and sometimes even hatred, for the poor, along with a 'winner-takes-all' mentality."[5] Winner takes all and winner controls all: one-tenth of 1 percent of Americans provided 40 percent of the campaign contributions during the 2016 presidential campaign—that is to say, 24,949 people.[6] These are the same people, or at least the same class of people, who have kept us from taking action on climate change. And so, we need to understand what drives them. We need to diagnose the intellectual and spiritual hookworm that has entered their bodies and attached itself to their brains.

The first thing to say is that current levels of inequality are almost beyond belief, deadly serious but also cartoonishly comical. The world's eight richest men possess more wealth than the bottom half of humanity. This trend, not surprisingly, is most pronounced in the United States, where the three richest men have more wealth than the bottom 150 million people taken together.[7] The richest tenth of 1 percent own about as much as the poorest 90 percent combined.[8] Jeff Bezos, pre-divorce the richest human, would have to spend $28 million every day just to keep his wealth from growing—which is funny in a sick way, given that in 2017 his median employee made $28,000 a year.[9] One family, the Waltons, of Walmart lineage, have more wealth than 42 percent of American families combined.[10]

The situation, of course, is worst for precisely whom you'd expect. The average black family in America has household wealth of $1,700, and that's been falling steadily.[11] And of course it's getting worse for each generation that comes along, as student debt increases and wages stagnate. (On the bright side, Credit Suisse reports a global uptick in billionaires under the age of forty; there were forty-six of them in 2017, up from just twenty-one in 2003.[12] And in the summer of 2018, *Forbes* reported that Kylie Jenner, youngest of the Kardashian clan, was about to become the forty-seventh, on the strength of her lip gloss line.) American children are now far less likely than members of previous generations to earn more than their parents, a pretty basic change in how we perceive the world.[13]

But even if we're the lucky ones earning more than our parents—even if we're winners—that inequality carries a real cost. In 2009 a pair of epidemiologists, Richard Wilkinson and Kate Pickett, studied a range of high-income countries. They found that as inequality rose, so did the number of prisoners and school dropouts, the rate of teen pregnancy and drug use, the incidence of mental illness and obesity. They compared a baby born in Greece with one born in America, where per capita income is twice as high and where we spend twice as much per person on medical care. But the Greek baby will live, on average, 1.2 years longer. "There are now many studies of income inequality and health that compare countries, American states, or other large regions, and the majority of these studies show that more egalitarian societies tend to be healthier," they observed.[14] It's an experiment that works for everything from literacy rates to murder rates. It's not, it turns out, because inequality hurts those at the bottom; it hurts *everyone* in a society.

In 2015, two statisticians pointed out another remarkable trend: death rates had begun to rise for middle-aged white Americans. The numbers were so unexpected that at first the researchers didn't believe their own work. "Mortality rates have been going down forever," said one of them, the Nobel Prize–winning researcher Angus Deaton. "There's been a huge increase in life expectancy and reduction in mortality over 100 years or more, and then for all of this to suddenly go into reverse, we thought it must be wrong. We spent weeks checking our numbers because we just couldn't believe that this could have happened." Convinced that the numbers were in fact correct, and that, indeed, the trend stretched back to 1999, Deaton and his research partner started looking for explanations. The hardest-hit groups were working-class Americans. So, in contrast with people in other parts of the world, they weren't as poor—they had televisions and cell phones—but they also had the anxiety that came from being near the bottom of a shifting economy in a country with a limited safety net. The researchers labeled this new mortality "deaths of despair." It comes from suicide and opioid abuse and bad nutrition, and it doesn't show a similar rise in the far more equal countries of Europe that, of course, were dealing with the same global recession, the same rise of China, the same hollowing out of manufacturing economies.[15] In 2016, Alan Krueger, the former chairman of the president's Council of Economic Advisers, released a study showing that half the prime working-age men who'd dropped out of the labor force were taking pain medication daily. They were also, on average, watching screens of one kind or another forty hours a week, as if it were a full-time job.[16] That's life for a great many people in the richest, and most unequal, society the world has ever built.

There's a tendency on the left to attribute all this—the inequality, the sanctioned greed, the environmental destruction—to "capitalism," and on some level, that's correct: the winners in this class war have used the mechanics of our economic system to gain their advantage. But it also lets the villains in this story off too easily. I mean, if you go to Scandinavia, you're in a recognizably capitalist world—there are markets, and you hand over money for goods, and some people get richer than others. (They're good at it, too. In the course of the twentieth century the Swedish stock market outperformed all others.) But it's a remarkably different flavor of capitalism. In fact, if you wanted, you could call it democratic socialism. It comes with high taxes and generous social services, with thorough government regulation and a commitment to rough equality. (Just to give the smallest example, Scandinavians sensibly base traffic fines on a person's annual income: a Finnish cell phone magnate recently paid $103,000 for riding his motorcycle the metric equivalent of forty-five miles per hour in a thirty-miles-per-hour zone.) The kind of capitalism turning America into a creepy jungle is very different: call it laissez-faire, or neoliberalism, or "getting government out of the way," or being "corporate-friendly." Whatever you call it, it's a particularly rapacious variant that's causing our current problem, one that's worth careful study.

Race, too, has clearly played an enormous role in driving our political transformation: It is the great stain on our nation, our irreducible sin. And there is no question that the wealthy used the hatreds and fears of too many white Americans as a massive lever to shift the country's politics. The Republican "Southern strategy" has been the spectacularly successful effort to play on deep reservoirs of racial hatred (in every region), and it has its echoes in the right-wing nationalisms of many other nations. It's particularly obvious right now, when an actual racist, Donald Trump, presides at the White House. And yet I think the men who really conceived and carried out that shift, from the Nixon era to the Reagan era to the present, were themselves driven less by white supremacy than by cynical calculation. They may have been racist in their intentions, and certainly in their effects, but they were also driven by a different kind of raw ideological energy.

This ideological energy needs to be understood. Something else has happened here, alongside profit-seeking and race-baiting. Some of us have come, in short order, to have a very different sense of what it means to be a person. Basic human solidarity has, especially for the most powerful among us, been replaced by a very different idea.

9

You could argue that the most important political philosopher of our time is the novelist Ayn Rand. Indeed, given the leverage of the present moment, leverage that is threatening to end the human game, you could argue that she's the most important philosopher of all time. She would have agreed: she once told a reporter that she was "the most creative thinker alive," and that the only other philosopher who'd influenced her was Aristotle.

At one level, that's nonsense. Rand might as well have written with a crayon; her ideas about the world are simple-minded, one-dimensional, and poisonous. But you don't need to be right to be influential. Her books animated many of the people who dominated American politics at crucial moments. When the United States was occupying the role of superpower, charting the course for a planet, she was occupying the hearts and minds of many of its most powerful people.

Consider Alan Greenspan, the avatar of neoliberalism and the chief architect of the world's economy in the years after the collapse of the Soviet Union—as *The Economist* once said, "the money men" regarded him as "Saint Alan."[1] He met Rand in the early 1950s, when he was a twenty-five-year-old economic forecaster. She was already the famous author of *The Fountainhead*, and he joined her circle, coming to her New York apartment every Saturday night to listen as she read drafts of her forthcoming novel, *Atlas Shrugged*, to her acolytes. When that book was published in 1957, the *Times* panned it, but Greenspan wrote a letter to the editor in its defense: "*Atlas Shrugged* is a celebration of life and happiness. Justice is unrelenting. Creative individuals and undeviating purpose and rationality achieve joy and fulfillment. Parasites who persistently avoid either purpose or reason perish as they should."[2] For the next decade, he published articles in Rand's magazine, *The Objectivist*, and when he was named by Gerald Ford to chair the president's Council of Economic Advisers, Rand stood beside him at his swearing in. She was dead by the time Ronald Reagan named Greenspan to chair the Federal Reserve, but thanks to people like Greenspan, her influence lived on—the Thatcher-Reagan years were, in the words of one writer, "the second age of Rand . . . when the laissez-faire philosophy went from the crankish obsession of right-wing economists to the governing credo of Anglo-American capitalism."[3]

And it was precisely America, in precisely those decades, that may have decided the planet's geological and technological future. Rand called her theory "objectivism," and usually it's grouped with "libertarianism," which at its most sophisticated and academic is a real economic school, with graphs and charts and equations. It has detailed theological arguments, some of them leading to sensible conclusions—that the war on drugs was a stupid failure, for instance, or that people should be able to marry whom they like—but its emotional core, channeled perfectly by Rand, is simple: Government is bad. Selfishness is good. Watch out for yourself. Solidarity is a trap. Taxes are theft. *You're not the boss of me.* When the *New York Times* described Rand as the "novelist laureate" of the Reagan administration, those were the kinds of raw attitudes it was referring to. Reagan's secretary of commerce, for instance, was a man named William Verity, who had inherited a steel fortune. He kept a card on his desk with a typically snarling quote from Rand: "There's nothing of any importance in life—except how well you do your work. Nothing. Only that. Whatever else you do will come from that. It's the only measure of human value. *All the codes of ethics they'll try to ram down your throat are just so much paper money put out by swindlers to fleece people.*"[4]

The legacy just kept getting stronger as the years rolled on past Reagan. Corporate barons and congressmen kept handing one another copies of Rand's novels: "I know from talking to a lot of Fortune 500 CEOs that 'Atlas Shrugged' has had a significant effect on their business decisions," the chief executive of one of America's largest regional banks said. "It offers something other books don't: the principles that apply to business and life in general. I would call it complete."[5] Rand's influence crosses oceans: the "intellectual architect of Brexit" keeps a photo of her on his desk.[6] Paul Ryan told the Atlas Society, a Rand fan club, that her books were "the reason I got involved in public service," and that he had required all his interns to read them. "I think Ayn Rand did the best job of anybody to build a moral case for capitalism," he explained in a series of videos posted to Facebook.[7] Rand was the inspiration for his "Path to Prosperity" budget, which called for ending Medicare. Yet Ryan also demonstrated the one problem with Rand for Republican politicians—she was, as one would expect from her views, vehemently opposed to Christianity, calling the Gospels "the best kindergarten of Communism possible." So, when Mitt Romney named Ryan as his vice-presidential running mate in 2012, the ever-courageous Ryan announced that he had long since "rejected" Randianism as "antithetical to my worldview."[8]

But he was almost alone in his apostasy. Clarence Thomas, before he was a Supreme Court justice, insisted that his staff at the Equal Employment Opportunity Commission watch the film adaptation of *The Fountainhead* over lunch; one aide called it a "sort of training film." In fact, when the White House was reviewing possible candidates for the federal bench, Thomas was recommended by the right-wing Institute for Justice precisely because of "his devotion to the philosophy of Ayn Rand."[9] Rex Tillerson, the former secretary of state, says *Atlas Shrugged* is his "favorite book." Ditto his successor, Mike Pompeo. Indeed, the billionaire Ray Dalio, one of those confidants Donald Trump calls late at night when he can't sleep, said, "Her books pretty well capture the mind-set" of the president and his men. "This new administration hates weak, unproductive, socialist people and policies and it admires strong, can-do profit-makers."[10] Andrew Puzder, Trump's first nominee for secretary of labor, named his private equity fund after Howard Roark, one of Rand's fictional heroes.

And what of the great man himself? Donald Trump has called *The Fountainhead* his favorite book. "It relates to business and beauty and life and inner emotions," he told *USA Today*. "That book relates to . . . everything."[11]

The cult of Ayn Rand extends far beyond the richest and most powerful. When the Modern Library asked readers in 1998 to catalogue the greatest books of the twentieth century, forget Hemingway and Joyce and Bellow: *Atlas Shrugged* and *The Fountainhead* were ranked one and two. Plenty of readers might have agreed with Barack Obama, who described Rand's work as "one of those things that a lot of us, when we were 17 or 18 and feeling misunderstood, we'd pick up."[12] But plenty of others have never put her down. One biographer described her as "the ultimate gateway drug to life on the right."[13]

When powerful people tell you, over and over, that these are the most important books they've ever read, that they shaped their thinking and bent their lives in a particular direction, when they love something so much that they name their hedge funds or their yachts in tribute, the rest of us should pay attention. So, why does Rand strike so many so hard?

Let's begin with what she got right. One way to think about her, and about right-wing laissez-faire neoliberalism in general, is as the toxic overshoot of a natural and appropriate reaction to the totalitarian threats of the blood-soaked twentieth century. The journalist Thomas Ricks recently published a fascinating joint biography of Winston Churchill and George Orwell, two very different men who, by the end of their lives, were united in agreement that the chief task facing humans was to "preserve a space for the individual in modern life" against the threat of the all-powerful state.[14] On the day that Britain declared war on Nazi Germany, Churchill said, "This is no war for domination or imperial aggrandizement or material gain. . . . It is a war, viewed in its inherent quality, to establish, on impregnable rocks, *the rights of the individual*, and it is a war to establish and revive the stature of man."[15] Though he fought the Second World War alongside Stalin, Churchill feared totalitarian communism at least as much as Nazism, just as Orwell, who had gone to Spain to fight fascism, ended up writing his greatest novels against a lightly disguised Soviet Union. As Orwell wrote toward the war's end, "This is the age of the totalitarian state, which does not and probably cannot allow the individual any freedom whatever. When one mentions totalitarianism, one thinks immediately of Germany, Russia, Italy, but I think one must face the risk that this phenomenon is going to be world-wide."[16]

Ayn Rand was working the same vein, and in her own case it was even more deeply imprinted—she didn't have to imagine what that totalitarian state might feel like. Born Alisa Zinovyevna Rosenbaum in 1905, she grew up in a Jewish middle-class household in Saint Petersburg. (Her best friend was Vladimir Nabokov's younger sister Olga.) In 1918, in the wake of the Bolshevik Revolution, members of the Red Guard pounded on the door of her father's successful pharmacy and told him it had been seized "in the name of the people." In the words of Rand's biographer Jennifer Burns, Alisa, "twelve at the time, burned with indignation. The shop was her father's; he had worked for it, studied long hours at university, dispensed valued advice and medicines to his customers. Now in an instant it was gone, taken to benefit nameless, faceless peasants." The soldiers had come carrying guns, threatening to kill her father, yet "they had spoken the language of fairness and equality, their goal to build a better society for all. Watching, listening, absorbing, Alisa knew one thing for certain: those who invoked such lofty ideals were not to be trusted. Talk about helping others was only a thin cover for force and power."[17]

The family fled toward Crimea, then under the control of the White Russians, but the Bolsheviks followed, and before long, the Rosenbaums' property was seized again, and they were reduced to selling off the family jewels to survive. Alisa made it to the university in what was now called Petrograd, surviving yet another purge of bourgeois students, and fell in love with Nietzsche and, especially, Aristotle. "Consistency was the principle that grabbed her attention," said Burns, "not surprisingly, given her unpredictable and frightening life."[18] She was living with her parents, but home was now a slum, and food was in short supply. Five million Russians starved to death in the famine of 1921–22, and city dwellers subsisted on ration cards. It is unsurprising that Alisa took the first opportunity to escape: she traveled to America on a short-term visa, ostensibly to visit family members in the Midwest, but even as she left, traveling under her new pen name, she knew she wasn't coming back.

Arriving in Chicago, she spent every possible moment in the movie theater. According to her journal, she watched (and meticulously ranked) 138 movies between February and August of 1926; her favorites were Cecil B. DeMille extravaganzas. She traveled to Hollywood, and in what sounds like a scene from a film, she saw the director behind the wheel of his idling car while he was speaking with someone. In Burns's recounting, Rand "stared and stared. DeMille, though used to adulation, was struck by the intensity of her gaze and called out to her from his open roadster. Rand stammered back in her guttural accent, telling him she had just arrived from Russia. DeMille knew a good story when he heard it and impulsively invited Rand into his car. He drove her through the streets of Hollywood, dropped famous names," and asked her to the set of *King of Kings* the next day. She parlayed the meeting into a job as a junior writer at his studio, summarizing properties that DeMille owned and suggesting how the scripts could be improved. And she went to work on her own screenplay, modeled on the sensational case of a teenage murderer named William Hickman, who had mutilated his victim and "boasted maniacally of his deed when caught." Rand was nonetheless sympathetic, even enraptured—to her, Hickman embodied "the strong individual breaking free from the ordinary run of humanity."[19] She quoted him in her journal: "What is good for me is right," adding her own response: "This is the best and strongest expression of a real man's psychology that I have heard."

And so, from the very beginning, we have the Rand who will eventually become famous. Here's what she wrote in a brief autobiographical sketch at the age of thirty: "If a life can have a theme song, and I believe every worthwhile one has, mine is a religion, an obsession, or a mania or all of these expressed in one word: *individualism*."[20] Given her early life, it made complete sense. Anyone of any spirit, watching at age twelve as his or her father is robbed at gunpoint, would hate the robbers; anyone with a spirited mind would be able to draw the broader conclusion about the system that perpetrated the robbery. But Rand, unlike an actual great thinker, could see no experience but her own, and her emotional need for consistency pushed her constantly to generalize from that experience. Had she devoted herself then to essays and manifestos, she would have been a minor and forgotten example of that twentieth-century type, a crank.

Instead, she wrote stories. And that made all the difference, because, of course, stories are how we understand the world. The fact that they were melodramas, the kind of writing that appeals to teenagers, or to those who don't read many books, would have been a wise tactical decision, though in the event, it seems to have simply reflected how Rand thought.

The Fountainhead tells the story of an architect, Howard Roark. He is the greatest architect on Earth, though of course no one recognizes this because everyone else is a bunch of collectivist "second-handers" who merely mimic the work of others from the past. The buildings Roark designed, by contrast, "were not Classical, they were not Gothic, they were not Renaissance. They were only Howard Roark."[21] Consider Roark's inner thoughts during a visit to a rock quarry:

> He looked at the granite. To be cut, he thought, and made into walls. He looked at a tree. To be split and made into rafters. He looked at a streak of rust on the stone and thought of iron ore under the ground. To be melted and to emerge as girders against the sky. These rocks, he thought, are here for me: waiting for the drill, the dynamite, and my voice; waiting to be split, ripped, pounded, reborn; waiting for the shape my hands will give them.[22]

Do you understand why Donald Trump identifies so dearly with him, this mighty Roark, who "had not made or sought a single friend on the campus"? Oh, and Roark's also more or less a rapist—the "love story" that runs through the book involves him dominating the beautiful Dominique, a "brutal portrayal of a conquest, an episode that left Dominique bruised, battered, and wanting more."[23] Rand offered "conflicting explanations for the sadomasochistic scene" that is "one of the most popular and controversial parts of the book," Burns notes. It isn't real rape, Rand once explained to a fan; it is "rape by engraved invitation."[24]

The book reaches its climax, however, not in the bedroom but in the courtroom, where Roark has to defend himself after he has blown up a housing project. Why? Because it wasn't built exactly the way he'd designed it. That's an insult to individualism, to the idea, as he explains to the jury, that the "creative faculty cannot be given or received, shared or borrowed. *It belongs to single, individual men.*" Roark goes on to explain that "the creator's concern is the conquest of nature," while "the parasite's concern is the conquest of men." While the former is exercising his complete independence, which "cannot be curbed, sacrificed, or subordinated to any consideration whatsoever," the latter is sucking up to "secure his ties with men in order to be fed. He declares that man exists in order to serve others. He preaches altruism."[25]

Altruism was perhaps the dirtiest word in Rand's lexicon. It's a "weapon of exploitation," Roark sneers, one that "reverses the base of mankind's moral principles. Men have been taught every precept that destroys the creator. Men have been taught dependence as a virtue." The jury acquits Roark, who goes on to build a really giant skyscraper, and book buyers rewarded Rand, who went on to write one more massive novel.

Atlas Shrugged was her magnum opus, set in a dystopian near-future when the government has managed to stifle business with too many regulations. As a result, the nation's most capable industrialists, thinkers, and inventors have gone on a strike organized by a hero named John Galt. They disappear to a sheltered valley in the Colorado mountains, where they "re-create a nineteenth-century world." The former head of an aircraft company is a hog farmer, and so on—the point is, these producers lead moral lives because they do not extract resources from others via taxes, but instead depend on their own talents and ingenuity to advance. As before, there is a woman. ("Her naked shoulder betrayed the fragility of the body under the black dress, and the pose made her most truly a woman. The proud strength became a challenge to someone's superior strength.") And as before, there is a long, tendentious speech, this time not to a courtroom but over a radio network that the industrialists have hacked in order to broadcast a seventy-page exaltation of the 1 percent. As Galt explains to a supposedly fascinated nation, "The man at the top of the intellectual pyramid contributes the most to all those below him, but gets nothing except his material payment, receiving no intellectual bonus from others to add to the value of his time. The man at the bottom who, left to himself, would starve in his hopeless ineptitude, contributes nothing to those above him but receives the bonus of all their brains. . . . Such is the pattern of exploitation for which you have damned the strong."[26]

Also a best seller, *Atlas Shrugged* nonetheless seemed to be swimming against the prevailing tide. It came out in 1957: Within a few years, Rachel Carson would publish *Silent Spring*, to far greater acclaim, stripping some of the shine off modernity. In Rand's Manhattan, the great urbanist Jane Jacobs was busy taking down Robert Moses, the Roark-like New York master-builder who listened to no one as he built highways where he pleased. As the writer Andrea Barnet pointed out recently, a whole cadre of remarkable women came to the fore in those years, from Carson and Jacobs to Betty Friedan and Jane Goodall, and what they shared was a reaction to the "strict hierarchies and separations" of the 1950s. What they saw, instead, were "entities and connections, the world as a holistic system. Instead of sweeping generalizations, they saw complexity and fine-grained detail. Instead of the world as an inert place, they saw movement and flow, evolution and process."[27] The 1960s were about to turn into a triumphant moment for those who believed that we *weren't* just individuals: the civil rights movement, and especially Lyndon Johnson's Great Society, were way stations on what seemed the road to greater human solidarity. The culture wars were under way, and you wouldn't have bet on Rand.

Especially given that she herself was floundering. She'd written her final, endless novel high on Benzedrine, and it eventually left her close to a nervous breakdown. She'd retreated inside a small circle of acolytes (Greenspan included), who met in her apartment to listen each week so she could read them new prose. She carried on an affair with her chief disciple, a man named Nathaniel Branden (who had changed his surname from "Blumenthal" to ally himself more closely with her), but only after informing his wife, who was also part of Rand's inner circle, about their plans. (When the wife nonetheless developed "persistent anxiety attacks," Rand helpfully developed "a new theory of 'emotionalism'" to "explain" the cuckolded woman's feelings.)[28] Meanwhile, deeply disturbed that John F. Kennedy had told Americans to "ask what you can do for your country," she proposed a book called *The Fascist New Frontier.* The publisher balked at that title, so it became *The Virtue of Selfishness*, but without the melodramatic plot lines of her fiction, her philosophical essays were inert. Rand grew ill. She'd of course kept smoking, despite the medical warnings, lecturing audiences on the "unscientific and irrational nature of the statistical evidence" linking tobacco and disease.[29] And when she contracted lung cancer, she of course refused to admit that she had been wrong. (After some initial balking on philosophical grounds, she did allow herself to be enrolled for Medicare and Social Security.) She died in 1982, with a six-foot-tall floral arrangement in the shape of a dollar sign standing by her grave.

By that time, Ronald Reagan was running the United States, and Margaret Thatcher had Britain in her iron grip, and the two embodied Rand's basic ideas with melodramatic power of their own. Reagan's most famous line was "The government is not the solution to our problem; government is the problem." Thatcher at her most strident sounded as if she *were* John Galt. "You know," she once said, as if it were the most obvious thing on Earth, "there's no such thing as society. There are individual men and women and there are families."* Those radical antigovernment ideas carried the day. Soon they seemed less radical, and eventually, they were mere conventional wisdom. They came with harsh attacks on labor unions and "entitlements" and anything else that reeked of human solidarity. At the moment of greatest leverage, they shaped America's choices when it was the most important country on the planet.

Rand had not done it by herself—as we shall see, there were other, far more systematic thinkers working the same ground, and far more diligent and effective political organizers—but she had told a story that made enough emotional sense to enough people at the top of the heap that it helped reshape the workings of her adopted nation.

And the reason it made enough sense was because it was rooted in something very real: that moment when the Red Guard seized her father's drugstore "in the name of the people." I visited the Soviet Union only when it was on its last legs, but that was enough for me to see that it had been a failure in every sense: its environment wrecked because

*Thatcher was nonetheless not a climate denier; as early as 1989 she told the United Nations that global warming was "real enough for us to make changes and sacrifices so we may not live at the expense of future generations."

people couldn't protest pollution, its people demoralized because they were told what to do with their lives, its official art and literature a dim-witted joke. Even the economy, whose success was supposed to justify all the repression, didn't work, because central planning on that scale turns out to be a stupendously bad idea. I remember standing with my wife outside Moscow's most prestigious retail outlet, the vast GUM department store. There were two long lines of people waiting, one outside the left door and one outside the right. When the left door opened first, everyone on the right side simply went home: they'd guessed wrong and knew that whatever was for sale would be long gone before they made it inside. Meanwhile, the lucky shoppers surged through the left door and past various empty departments and up the stairs into the one busy section, which was selling winter coats for children. You had to show your residence certificate even to have a chance of buying one. An industrial society that can't produce enough children's coats for the Russian winter—that's failure on a grand scale.

The great advantage of the twenty-first century should be that we can learn from having lived through the failures of the twentieth. We're able, as people were not a hundred years ago, to scratch some ideas off the list. It's very useful to know, for example, that state communism is a really terrible idea. And it's not as if we have nothing any longer to fear from a powerful central government; as Edward Snowden demonstrated, nation-states are still in the surveillance business. Also, Alexa.

But overlearning the lessons of the past is just as dangerous as ignoring them. If you can't distinguish between national health insurance and indentured servitude, if Denmark reminds you of North Korea, then you damage the present in the name of the past. If you must resist the Clean Air Act because of your visceral fear that it might lead to so much government that your drugstore gets taken away, then the twentieth century has become not a teacher but an irrational barrier.

Unless, of course, you make a fortune violating the Clean Air Act.

10

Ayn Rand's father had his door kicked down by the Red Guard in 1918.

Ninety-six years later, in the fall of 2014, Tom Perkins feared he might be the next victim of a purge—a "Kristallnacht," he called it, in a letter to the *Wall Street Journal*. Perkins was one of the richest men on the planet. He had literally built one of the largest yachts on earth, the 289-foot *Maltese Falcon* (which carried its own submarine, which he named the *Dr. No*). But Perkins had noticed that the poorer people of San Francisco were demonstrating against the real estate takeover of their city by tech barons and venture capitalists like him—indeed, some unruly hooligans had dared to smash a piñata in the shape of one of the swank buses Google uses to ferry its workers from their urban lofts out to Silicon Valley. And so, Perkins thoughtfully wrote to the *Journal* to "call attention to the parallels of fascist Nazi Germany in its war on the 'one percent' namely its Jews, to the progressive war on the American one percent, namely the 'rich.'" On television the next day, he explained that he simply wanted to make sure that Americans didn't "demonize the most creative part of society." After explaining that the watch he was wearing was worth "a six-pack of Rolexes," he added, "We are beginning to engage in class warfare. The rich as a class are threatened through higher taxes, more regulation."[1]

It's easy to ridicule Tom Perkins. (In fact, it's so easy that one can't resist. Among other things, he once wrote a novel called *Sex and the Single Zillionaire*, about a wealthy widower named Steven Hudson who is asked to be on a reality show where "a gaggle of gold-digging bimbos" compete for his love. Many of them eventually end up back in his marvelous Manhattan penthouse, "a sleek ultra-modern, minimalist yet somehow very comfortable, glassed and terraced condominium." And once they get there, well: "Heather's body was glistening with perspiration as she moaned in anticipation of the whiplash. . . .")[2] As I was saying, it's easy to make fun of Tom Perkins, but in fact, the tiny group of men who are his economic peers have come to dominate our political life, making precisely the choices that may cut short the human game. And their language is always the same: the "producers," the "creators," the "people of value" are threatened by the mob. And so, they must organize and fight. Most of them aren't quite as blatant or as public as Perkins, who a month after his "Kristallnacht" letter told an audience that, in his ideal world, "you don't get the vote if you don't pay a dollar in taxes. But what I really think is it should be like a corporation. You pay a million dollars, you get a million votes."[3] Still, in essence, they're thinking just like him.

Just to be clear: I'm arguing that a systematic idea about the world emerged in the latter half of the twentieth century, an idea as potent in its way as Leninism had been in the first half. This idea (that government was bad, and that productive individuals and their corporations needed to be freed from its clutches) changed the politics of America, changed it enough that even liberals couldn't or wouldn't stand up to it: Bill Clinton, swimming with this current, managed to force through Congress both NAFTA and the General Agreement on Tariffs and Trade, enshrining unrestricted global trade as the obvious goal. He also ended "welfare as we know it," helping build the each-to-his-own-devices nation we now inhabit. This is what happens when someone manages to change the zeitgeist; this is why Rand and Reagan were so crucially important.

The Koch brothers, Charles and David, are the classic exemplars of this worldview, and arguably the most powerful men in the Western world. They're not as blustering as Donald Trump, and they're uninterested in his flashiest hatreds and crusades, but they're both the most important architects, and among the biggest beneficiaries, of his rule. On everything from tax cuts to environmental regulation, the Trump years, for the Kochs, have been what their biographer Jane Mayer calls "their dream come true."[4]

The Koch brothers have become such a shorthand for plutocratic excess that it's important to remind ourselves that they are real men with real stories, also rooted in the twentieth century (and told most ably by Mayer in her book *Dark Money*). Not unlike some Ayn Rand hero, their father, Fred Koch, had invented an improved process for refining crude oil into gasoline. The Soviet Union sought his expertise as it set up its own refineries after the Bolshevik Revolution. At first, Fred said he didn't want to work for Communists, but because they were willing to pay in advance, he overcame his scruples and helped Stalin meet his first five-year plan by building fifteen refineries and then advising on a hundred more. Next, Fred turned to another autocrat with busy expansion plans, Adolf Hitler, traveling frequently to Germany, where he "provided the engineering plans and began overseeing the construction of a massive oil refinery owned by a company on the Elbe River in Hamburg," in Mayer's description. It turned into a crucial part of the Reich's military might, "one of the few refineries in Germany" that could produce "the high-octane gasoline needed to fuel fighter planes."[5] And it turned the elder Koch into an admirer of the regime, one who, as late as 1938, was writing to a friend that "the only sound countries in the world are Germany, Italy, and Japan, simply because they are all working and working hard." Comparing the scenes he saw in Hamburg to FDR's New Deal, Fred Koch said it gave him hope that "*perhaps this course of idleness, feeding at the public trough, dependence on government, etc., with which we are afflicted is not permanent and can be overcome.*"[6]

Fred met his wife, Mary, at a polo match in 1932, when his "work for Stalin had put him well on his way to becoming exceedingly wealthy." They built a Gothic-style stone mansion on the outskirts of Wichita, Kansas, with stables, a kennel for hunting dogs, and the other paraphernalia required for pretend gentry, and in the first eight years of their marriage, they had four sons: Frederick, Charles, and a pair of twins, David and William. The first two were raised by a German nanny, who "enforced a rigid toilet-training regimen requiring the boys to produce morning bowel movements precisely on schedule or be force-fed castor oil and subjected to enemas." Luckily for the twins, she left for home when they were born, apparently because "she was so overcome with joy when Hitler invaded France she felt she had to go back to the fatherland in order to join the führer in celebration."[7]

Of those four sons, Charles became the dominant force, and David his close colleague. Eventually, by Mayer's account, they tried to blackmail the eldest brother, Frederick, out of his share of the family business by threatening to tell their father that he was gay. Bill, too, later parted ways with his brothers, parlaying his share of the inheritance into a lucrative oil business and then using some of the proceeds to fund opposition to wind turbines off his Cape Cod beach. But Charles was always the crucial Koch, the one who followed most closely in their father's wake.

Fred, despite the original source of his fortune, had become a fervent anticommunist and one of the eleven founding members of the John Birch Society. One of the figures in that orbit, Robert LeFevre, became Charles's original guru, opening a "Freedom School" in Colorado Springs in 1957, where he preached not just the Birchers' anticommunism but also an adamant opposition to America's government. "Government is a disease masquerading as its own cure," LeFevre insisted, and by 1966, Charles was a trustee of the school, where he celebrated the work of the Austrian economists Ludwig von Mises and Friedrich Hayek and, through them, the world of a libertarianism more orthodox (and less melodramatic) than Rand's. Many of the disciples of that so-called Mont Pelerin movement (named for the Swiss resort where the libertarian luminaries first gathered in 1947) would go on to hold high office, constructing the basic neoliberal economic framework we've lived under since Reagan.

But Charles Koch was younger and fiercer still, intent on truly revolutionary change. By the mid-1970s, he'd founded the Center for Libertarian Studies in New York and written a paper "on how the fringe movement could obtain genuine power." His essay was notable for, among other things, its endorsement of secrecy. "In order to avoid undesirable criticism," he wrote, the details of "how the organization is controlled and directed should not be widely advertised."[8] At first, he worked through the Libertarian Party, persuading his brother David to be its vice-presidential candidate in 1980 because, that way, they could use their own money and avoid campaign finance laws. But the ticket's poor showing (only 1 percent of the vote) convinced him they needed to work behind the scenes, using the Republican Party as their vehicle. Reagan had started the Earth moving in a new direction, but after decades of the New Deal and the Great Society, he could only begin the job. The Kochs wanted to go much farther; they wanted a full-on quake.

Enter a Southern academic named James McGill Buchanan, who provided Charles Koch and others with, in the words of historian Nancy MacLean, "an operational strategy to vanquish the model of government they had been criticizing for decades." Buchanan was a fairly obscure economist who, in the late 1950s, began setting up a series of (well-funded) university research centers to "train a new line of thinkers" who would counter those seeking to impose "an increasing role of government in economic and social life" and would, instead, stand for a "social order . . . built on individual liberty."[9] Buchanan's particular contribution, which won him a Nobel Prize, was an economic thesis that came to be called public choice theory. It held that "allocating resources by majority decision-making invited voters to group together as 'special interests' or 'pressure groups' in collective pursuit of 'profits' from government programs." That is to say, Buchanan thought that majorities wanted "free stuff" from governments and would vote for politicians who obliged them, and that those politicians would, in turn, tax the rich to pay for it all. The result was "an overinvestment in the public sector" because this powerful coalition of voters, politicians, and bureaucrats (who liked the government jobs that resulted) could foist the cost on to the billionaire "victims" of excess taxation. This overinvestment, in turn, held down capital accumulation, and hence investment, and hence economic growth.[10]

All that sounds, at first blush, like fairly standard Republican boilerplate, the familiar attack on "big government." It's suspect in purely economic terms, as many analysts have pointed out: economies actually do well when we invest in things such as education and roads. But if you think about it carefully, as both Buchanan and the Kochs did, you realize there's another reason the analysis is unlikely to prevail. And that reason is . . . democracy.

Mitt Romney discovered this particular flaw on the day in 2012 when he told attendees at a high-dollar fund-raising event that roughly half of Americans would vote to reelect Obama simply because they were moochers. A waiter with a hidden camera captured his words: "Here are 47 percent who are with [Obama], who are dependent upon government, who believe that they are victims, who believe the government has a responsibility to care for them, who believe that they are entitled to health care, to food, to housing, to you name it. That that's an entitlement. And the government should give it to them. And they will vote for this president no matter what." Americans were shocked to hear the Republican presidential nominee state it so bluntly—this was the day Romney definitively lost the election—but only because they hadn't been paying much attention to what the libertarian right really believed. When Romney went on to say, "I'll never convince them that they should take personal responsibility and care for their lives," he was merely echoing the rhetoric that men and women such as Buchanan, Koch, and Rand had been developing for decades. This belief, that they were being unfairly taxed to support the lazy, was at the core not just of their politics but of their emotional worldview. When Charles Koch decided to get married, he insisted that his wife be "indoctrinated with these ideas, lest their marriage lack harmony of purpose." The wedding couldn't take place until this "intense training" had succeeded, which apparently didn't take too long: Elizabeth Koch was soon complaining that America had become "a country of non-risk-takers," the sort of people "who just want to be coddled and taken care of."[11]

However appealing that logic sounds to the very rich, it offends far more people. In a fair fight—that is, in a working democracy—billionaires won't win most of their battles because most people will vote for their own interests. So, the billionaire conclusion was, in essence, that democracy couldn't be allowed to really work—that's what Tom "Biggest Yacht" Perkins meant when he suggested the rich should get more votes. Such straightforward remarks were better confined to private talks and obscure journals, but this was the central premise of what became a concerted effort to undermine democratic institutions. Buchanan once asked, "Why must the rich be made to suffer" as a result of "simple majority voting"? A citizen "who finds that he must, on fear of punishment, pay taxes for public goods in excess of the amount that he might voluntarily contribute" is no different from someone mugged by a "thug who takes his wallet in Central Park." He advocated a "generalized rewriting of the social contract," giving America "a new set of checks and balances," changes that would be "sufficiently dramatic to warrant the label 'revolutionary.'"[12]

And so, they went to work, quietly building the networks and the institutions that could deliver that change. The Kochs set up a web of political operatives in groups such as Americans for Prosperity. They and their allies also erected think tanks and academic centers around the nation that churned out policy papers to back up their plans and to produce "messaging" to convince people to vote against their own interests. (The classic examples included recasting the inheritance tax, paid by a tiny slice of the 1 percent, as a "death tax" that all should fear.) Groups such as the Federalist Society concentrated on the judicial branch, vetting candidates who would uphold libertarian ideas. Fox

News, America's first partisan television network, emerged to amplify the stream of messages. Much of the work was done at the state and local level, where local billionaires—junior Kochs, as it were—concentrated on winning statehouses, gerrymandering districts, and enacting the voter-suppression laws that reduced the size of the "mob" they'd need to vanquish.

Most of all, they used the main thing billionaires possess in surplus: money. As Karl Rove said shortly after the Supreme Court ruled in *Citizens United*, "people call us a vast right-wing conspiracy, but we're really a half-assed right-wing conspiracy. Now it's time to get serious." In 2010, "Republican-aligned independent groups" spent a completely unprecedented $200 million on midterm elections, gaining 63 seats in the House. In 2012, more than 60 percent of all campaign contributions came from less than one-half of 1 percent of the population. In 2016, the Koch networks vowed to spend a completely absurd $889 million on the election. "We've had money in the past, but this is so far beyond what anyone has thought of it's mind-boggling," the head of Common Cause said. "This is unheard of in the history of the country. There has never been anything that approaches this."[13] As it turns out, the Koch operation spent "only" about three quarters of a billion dollars, mostly on congressional and state races, because, for a moment, Trump seemed to be upending their calculations. For all his love of Ayn Rand, he ran what sounded more like a Democratic campaign: on health care, for instance, he said, "We're going to have insurance for everybody. . . . There was a philosophy in some circles that if you can't pay for it, you don't get it. That's not going to happen with us." In fact, he promised, "everybody's going to be taken care of much better than they're taken care of now." But as we now know, Trump was simply bringing something new to the game: not clever messaging, but brazen lying.

And arriving in Washington with no existing ideology except feeding his narcissism and enriching his family, Trump proved the perfect president finally to enact the full government-hating agenda. Robert Mercer, who'd funded not only Trump's campaign but also Cambridge

Analytica, the source of so much Facebook skullduggery, was a key figure—and a classic Randian. As one colleague explained, "Bob believes that human beings have no inherent value other than how much money they make. A cat has value, he's said, because it provides pleasure to humans. But if someone is on welfare[,] he has negative value. [Bob] thinks society is upside down—that government helps the weak people get strong, and makes the strong people weak by taking their money away, through taxes." Another colleague explained of Mercer, "He thinks the less government the better. And if the president's a bozo? He's fine with that. He wants it to *all* fall down."[14]

In the meantime, Mercer and his pals were happy to assist with reorienting the Trump administration to serve their goals. "The vacuum in Trump not having his own network is filled by people who've been cultivated for years by the Koch network," said one expert.[15] Marc Short, the former director of the Kochs' Freedom Partners investment fund, became Trump's legislative director; Short's major triumph was the huge tax cut for the rich that the Senate passed at the end of 2017. The win required massive lobbying from the Kochs' network of donors—"the most significant federal effort we've ever taken on," said the head of Americans for Prosperity.[16] The bill will save the Koch brothers and their company more than a billion dollars a year. Days after it passed, they donated $500,000 to Speaker of the House Paul Ryan's war chest, which actually seems like a pretty miserly tip. In 2018 they sent a memo to their wealthy donors taking credit for dozens of policy triumphs, from ending new overtime reforms to making sure that the discovery of Native American cultural artifacts didn't slow down oil drilling. It's true they didn't much like Trump, and eventually, given his narcissism, the sentiment was returned. By August 2018, the president was tweeting that, because the Koch brothers opposed his wall and his tariffs, they had "become a total joke in real Republican circles." But the joke seemed more likely to be on Trump. It was increasingly clear that the Kochs had milked the administration for what they really wanted (tax cuts, deregulation, Supreme Court justices) and were now looking forward to the Pence years.

And their work was by no means confined to Washington. Many state governments have been transformed by the Koch-funded American Legislative Exchange Council. Meanwhile, in cities and towns across America, the Kochs ran expensive campaigns opposing projects such as public transit, both because it dried up demand for the gasoline they sold and because buses and trains "go against the liberties Americans hold dear." As a spokesman for Americans for Prosperity patiently explained, "If someone has the freedom to go where they want, do what they want, they're not going to choose public transit."[17] Especially if there isn't any.

"We've made more progress in the last five years than I had in the last 50," Charles Koch told his fellow billionaires in 2017. "The capabilities we have now can take us to a whole new level."[18] And so they vowed to fight on—as the 2018 midterm election entered its final stretch, more than a quarter of the outside commercials in congressional races were coming from just two Koch-sponsored advocacy groups,[19] and the Kochs were also funding a "seven-figure ad buy" on behalf of the Supreme Court nominee Brett Kavanaugh, apparently pleased that his vote promised what one scholar called "the end of the regulatory state as we know it."[20] Their to-do list included yet more "reforms" of labor laws and, above all, reductions in "entitlement" spending such as Social Security, spending that, of course, they argued America could no longer afford because of the deficits left from those massive tax cuts.

Basically, they'd won.

11

But that win is not likely to last forever. Few wins do, and this one is particularly shaky because it rests on a fundamentally flawed sense of who human beings actually are. The idea that we are only individuals, that "there is no such thing as society," that we owe each other nothing—none of that fits with our deepest nature.

We are social creatures. I'm not going to bother with a potted history of our development as a species: cooperating to hunt game, developing language to make the hunt more successful. Suffice it to say that we evolved from the tribe, the band. Even Ayn Rand knew that, though she put her own nasty spin on it: "Civilization is the progress toward a society of privacy," she wrote. "The savage's whole existence is public, ruled by the laws of his tribe. Civilization is the process of setting man free from men."[1]

Setting us free from our fellow human beings is a profound mistake, because we haven't in fact evolved into some new creature. You can prove it easily: find an American who does not belong to any group—sadly, not that hard a task—and convince him to join a club or a society. The mere act of joining in with others *halves one's risk of dying in the next year.* Recent research makes clear that social separation damages us: individuals with fewer connections have disrupted sleep patterns, altered immune systems, and higher levels of stress hormones.

Isolated people have a 29 percent higher risk of heart disease and a 32 percent higher rate of strokes. It begins early: "socially isolated children have significantly poorer health twenty years later." (Even Rand made sure she lived life within a small circle of acolytes, her adoring tribe.) All told, loneliness is as bad for you physically as obesity or smoking.[2] It works the other way, too: "Study after study shows that good social relationships are the strongest, most consistent predictor there is of a happy life, even going so far as to call them a 'necessary condition of happiness.' . . . This is a finding that cuts across race, age, gender, income and social class so overwhelmingly that it dwarfs any other factor," reports the journalist and cultural critic Ruth Whippman.[3] Hell, it cuts across species: *ants* that are allowed to socialize live ten times longer than their isolated counterparts.

Altruism, what Rand called "the poison of death in the blood of western civilization," turns out to be a tonic instead. When neuroscientists study our brains, they find they "behave differently during an act of generosity than during a hedonistic activity," reports the *New York Times.* Dr. Richard Davison, founder of the Center for Healthy Minds at the University of Wisconsin, put it like this: "When we do things for ourselves, those experiences of positive emotions are more fleeting. . . . When we engage in acts of generosity, those experiences of positive emotion may be more enduring and outlast the specific episode in which we are engaged." Not surprisingly, then, older adults who volunteer to help children with reading and writing tend to experience less memory loss—that is, the great personal terror for most of us, losing our sense of self, becomes less likely if we engage with others.[4] Why does hearing loss increase your risk of dementia? Likely because hearing loss "tends to cause some people to withdraw from conversations and participate less in activities. As a result, you become less social and less engaged," according to the Cleveland Clinic.[5]

Even in corporate life, it turns out, those who "offer assistance, share valuable knowledge, or make valuable introductions" turn out to be far more useful for a business than those who "try to get other people to serve their ends while carefully guarding their own expertise and time," according to the *Harvard Business Review*. In thirty-eight studies across 3,500 businesses, the journal found that "higher rates of giving were predictive of higher unit profitability, productivity, efficiency, and customer satisfaction, along with lower costs and turnover rates."[6] That's square in the heart of the capitalist hurly-burly. Indeed, the Nobel Prize for Economics, once awarded to James Buchanan for his theory that most of us are parasites, was more recently awarded to Elinor Ostrom, the great theorist of the commons. What she found, across societies and historical eras, was that communities were quite capable of common-sense cooperation—the "tragedy of the commons" was usually not a tragedy at all, as long as no one decided they had to have it all. From lobster fisheries in Maine to irrigation systems in Spain to forests in Nepal, Ostrom found that "the schemes were mutual and reciprocal and many had worked well for centuries."[7]

It's not that the hyperindividualists are wrong. It's that they're half-wrong, and that's what makes them so dangerous. We *do* aspire to a certain amount of what Rand called "privacy." If you spend time in, say, rural China, you will visit plenty of homes where many members of a large family share a home or small compound. People will often sleep many to a room, and sometimes the family pig will sleep there, too. In that world, building an extra room so that a husband and wife can sometimes be alone—that's worth a lot. The economic literature on happiness makes it clear that up to a certain point, more income equals more liberation of this kind, and hence more satisfaction: one can leave one's village on a trip, which for most of human history was almost impossible. But the literature also makes it clear that past a certain, and surprisingly low, level, there's not much linkage between more money and more happiness. And in part that may be because money leads to *too much* "privacy." Postwar America spent most of its fortune on a single project, building bigger houses farther apart from one another, and the result was that people ran into each other less: the number of close friends claimed by the average American dropped by half.[8] Now we're pursuing the same project with our array of screens, as the psychologist Jean Twenge points out in her recent statistical portrait of young people currently in high school and college, whom she calls "iGen." These kids spend far less time hanging out with friends than any generation in history, and the data show they're uniquely unhappy as a result.[9]

There's actually a kind of natural balance between public and private that wise people have always recognized. Adam Smith can be said to have launched the movement that ended up with the Kochs: his landmark work, *The Wealth of Nations*, provided the first explanation of how pursuing one's own interest could end up increasing the general prosperity. But that wasn't Smith's only book. In *The Theory of Moral Sentiments*, he points out that "how selfish soever man may be supposed, there are evidently some principles in his nature, which interest him in the fortune of others, and render their happiness necessary to him." Self-interest was not the most admirable of our traits. Instead, Smith listed "humanity, justice, generosity, and public spirit . . . the qualities most useful to others."[10] But the economic tradition that grew up in his wake largely scorns those insights. Because markets have proved to perform so brilliantly at their particular task of creating wealth, economists have largely forgotten that there *are* other tasks.

In truth, this crudeness seems to affect most deeply those who study economics: researchers found that third-year students in economics rated altruistic values such as helpfulness, honesty, and loyalty as far less important than freshmen did. "After taking a course in economic game theory, college students behaved more selfishly and expected others to do so as well," they observed. And economics professors, it turns out, "give significantly less money to charity than their worse-paid colleagues in many other disciplines."[11] This is the world where think tanks debate whether it's cost-effective to save the Arctic, and where the *Wall Street Journal* runs a headline such as HOW DO YOU PRICE A PROBLEM LIKE

KOREA: ANALYSTS ARE TRYING TO WORK OUT WHAT HAPPENS TO MARKETS IN THE EVENT OF AN ALL-OUT NUCLEAR WAR. (In case you're wondering: in the event of a "potentially uncontained military conflict in which the global superpowers get involved," the yield curve on Eurobonds would "likely flatten due to weaker risk appetite.")[12]

Because this is so contrary to our nature, eventually even the U.S. political system will work its way back to some kind of balance. The Koch brothers may well have hit their zenith. Political scientists crunching the polling data said that the Kochs' two signature laws (the attempted repeal of Obamacare and the successful tax "reform" package) were the "most unpopular pieces of major domestic legislation of the past quarter-century," the journalist Michael Tomasky points out. Of the nine most popular recent laws, he observes, "eight pursued what could broadly be defined as liberal goals, like gun control and environmental protection."[13] For the last few years, America's most liked politician, by far, has been a socialist, Bernie Sanders, who campaigned on the antilibertarian slogan "Not Me, Us," and who holds up Scandinavia as a model. Denmark and Sweden and Norway, of course, are what this "balance" I've been describing looks like in practice: a market system with a strong commitment to social justice, the lowest levels of inequality on the planet, and, by most measures, the happiest citizens, people leading private lives, but not leaving others behind. You'd have to look hard to find a case of hookworm in Bergen or Aalborg.

The road back to balance will be long, and many people will suffer unnecessarily along the way, but the Randian view of the world is simply too disconnected from human nature to dominate us forever. Perhaps our electoral systems are strong enough to reverse the craziness: the midterm elections of 2018 represent a good start. (In early 2019 newly minted Congresswoman Alexandria Ocasio-Cortez called for nearly doubling the top U.S. tax rate on income over $10 million, and lots of people applauded.) Perhaps the power of money in our political life is so great that it will require something that looks more like a nonviolent revolution. But half lies have half-lives. Humans will rise to the occasion, in America and in all the other places that have, for the moment, been knocked off kilter.

Except there's a problem, a big one. There's too much leverage in the system.

In the past, when the ideological pendulum swung hard in one direction, there was time and space for it, eventually, to swing back. The Gilded Age robber barons (or, if you'd prefer, the captains of industry) pushed wealth, and hence political power, as sharply their way in the late nineteenth century as have the libertarian billionaires of our time. Statistics are harder to come by, but the estimate is that four thousand families in the 1890s had as much wealth as the other eleven million households in America. And so, that inevitably gave rise to the Populist and Progressive movements, and the income tax; inequality abated during the Progressive Era, only to rise again during the Roaring Twenties, only to fall sharply with the New Deal, the Second World War, and the mass prosperity that followed. Much harm was done along the way, but none of it permanent, at least in the largest sense—certainly none of it threatened to end the human game, *not because the robber barons were less venal, but because they lacked sufficient leverage* to make change on that scale, and because people stood up to them.

Or think about the Second World War—it was the greatest military conflagration the world had ever seen, with millions dead and whole continents turned upside down, but eventually the tank tracks eroded back into the landscape. Farmers still hit the occasional unexploded shell with their plows, but the world moved on, because there wasn't quite enough leverage to knock it entirely off its course. We were lucky—Hitler came close to developing nuclear weapons. Whether because of Heisenberg's treachery or simple incompetence, the Nazis didn't quite get there, but given another a year or two? We know how *we* would have reacted, because we dropped the bomb on Japan though the war was nearly won. Nuclear war with Germany might have been leverage enough to fundamentally and permanently alter the globe.

Global warming turns out to be the perfect example of too much leverage. The men who gained ideological power beginning in the Reagan years, a great many of them directly connected to the oil and gas industry, were in control at precisely the moment when they could do the most damage. In the years since 1990—the years since, say, the Exxons and Kochs of the world started launching the various "think tanks" and front groups to poison the debate with what they knew was a series of lies—the world has emitted more carbon dioxide than in all the decades before. And this turned out to be the crucial carbon dioxide. We know now that 350 parts per million carbon dioxide is the most we could safely have in the atmosphere, a number we've rocketed past in precisely those years. It didn't necessarily need to be that way. In a world with slightly different physics, 800 parts per million might have been the breaking point—in which case, we'd still have room to recover. If someone grabs the steering wheel when you're a mile from the cliff, you have time to wrestle it back. But as it turned out, we were on the edge of the abyss.

This leverage doesn't just guarantee a shift in climate; it also locks in new forms of inequality that can't be undone even by revolution. As the temperature climbs, it's the poorest who suffer most, a suffering that isn't going away. When the peak temperature in leafy suburbs can be lower by as much as fifteen degrees, "landscape is a predictor for morbidity in heatwaves," in the words of one study, which found that African Americans were "52% more likely than white people to live in areas of unnatural 'heat risk–related land cover.'"[14] Imagine what it's like in a refugee camp, or a prison. It's hell, is what it is.

And again, the people who were telling the lies knew they were lies. This wasn't a hard conspiracy to organize—just one hundred firms in the fossil fuel industry account for 70 percent of the planet's emissions. But it wasn't based on simple greed, either. Self-interest mixed perfectly with ideology. Remember the CEOs gleefully handing one another copies of *Atlas Shrugged* and the billionaires who had grown up in the fever swamps of the antigovernment movement? These guys thought they had cracked the code of history. Climate change was, for them, inconceivable because it would get in the way of profits—the Koch brothers run enormous pipeline networks; they are among the biggest leaseholders in Canada's tar sands—*but also because it marred the purity of their belief system*. The antigovernment forces had, at some level, no choice but to deny global warming, because tackling it would have required governments to take strong action—at the very least, to set a price on carbon so that markets could then work their putative magic. They believed more strongly in their particular economic fantasia than they did in physics or chemistry, and so they churned out an endless series of deceptions.

Rupert Murdoch, the planet's dominant communicator, provides a case in point because for a moment he appeared to be an exception to the rule. In 2007, in the wake of popular acclaim for Al Gore's film *An Inconvenient Truth*, he actually announced that he'd found climate religion. In a speech that he reprinted in his newspapers, he said, "Climate change poses catastrophic threats" and "[W]e can't afford the risk of inaction." He pledged to make NewsCorp carbon neutral, promising everything from electric golf carts on the 21st Century Fox studio lot to "the latest LED lighting technology in the master control rooms at Fox News." His outlets, he said, would "change the way the public thinks about these issues." Reality seemed to have broken through.[15] But that reality was always at war with Murdoch's deeper Randian ideology, the idea that government was out to steal from him. "What's fair about taking money from people who earned it and giving it to people who didn't?" he wrote a few years later.[16] This rhetoric sounded more authentic, and so it was probably inevitable that his pledge to spread the word about climate change would give way to lying about it in the service of the larger antigovernment cause. Between 2012 and 2016, even as the scientific case for climate danger became ever more obvious, Murdoch's *Wall Street Journal* published 303 op-eds, columns, and editorials on climate change, 287 of which were "misleading and debunked denial talking points, conspiracy theories, and political attacks." Or, put another way, roughly 95 percent of what he published "disagreed with the roughly 97 percent consensus among climate scientists."[17] In 2016, the hottest year ever recorded on our planet, Fox News devoted six minutes to covering the issue,[18] which may have been just as well, given that when the network's anchors did mention it, they were likely to say things like "No one is dying from climate change."[19] By the summer of 2018, even as the world experienced record heat and fire, the *Journal* was concluding, in proper antigovernment fashion, that "climate change is over. All that remains is . . . bureaucratic mandates on behalf of special-interest renewable-energy rent-seekers."[20]

The chair of the economics department at George Mason University, where James Buchanan ended his days in a Koch-funded sinecure, explained it perfectly: "Those of us who recognize these important benefits of capitalism . . . are reluctant to yield power to governments to tackle global warming." Indeed, he continued, perhaps "the best policy regarding global warming is to neglect it."[21]

This ideological *idée fixe* explains the single most bizarre talking point of the standard climate denier: the idea that climate scientists are "in it for the grant money" and hence skewing their results in a conspiracy to whip up anxiety and produce more government funding. The claim is obviously absurd, at least to anyone who has had the opportunity to meet climate scientists and oil executives and compare their lifestyles, but it draws from this deep well of suspicion about anyone who works for the government. In the early 1990s, scholars at the Center for the Study of Public Choice at George Mason wrote a book called *The Economics of Smoking*, which argues that because a cure for cancer would "put many anti-cancer bureaucrats out of work," these scientists had weak "incentives to find and develop effective treatments" and strong "incentives to magnify the risk of cancer."[22] Another of Buchanan's acolytes explained that "health officials' interest in testing small children's blood for lead made sense when one considered that finding poisoned children validated their jobs."[23] In other words, who would worry about lead poisoning in children if they weren't getting paid for it? (Probably not the key billionaire member of the Koch donor network whose company was found marketing arsenic-tainted mulch as "ideal for playgrounds.")[24] Believing that government doctors care about kids only in order to make more money, or that climate scientists pursue their work with profit in mind, is such a cynical view that it can be explained only as a reflection of life in the billionaire bubble. If greed warps your life, you assume it must warp everyone's.

So: global warming is the ultimate problem for oil companies because oil causes it, and it's the ultimate problem for government haters because without government intervention, you can't solve it. Those twin existential threats, to cash and to worldview, meant that there was never any shortage of resources for the task of denying climate change.

We've already seen how it began in the late 1980s, with Exxon-funded groups such as the Global Climate Coalition, but by the new century, it had morphed into a massive web of plutocrats and their functionaries. Koch-funded professors at four hundred colleges across America were busy teaching the new gospel. "Only idiocy would conclude that mankind's capacity to change the climate is more powerful than the forces of nature," one explained to his charges in Maryland. In Colorado, another produced and starred in the slyly titled film *An Inconvenient Truth . . . or Convenient Fiction?* At the University of Kansas, the former chief economist for one of the Kochs' enterprises took over the new Center for Applied Economics (funded by the Kochs) and went to work trying to repeal the state's "renewable portfolio standard," which mandated the construction of at least a little solar and wind power.[25] The Kochs' Kansas network also flew in a climate change–denying scientist who had received $230,000 in grants from the Charles G. Koch Foundation (and who explained in one memo that he was in the business of creating "receivables" for his fossil fuel industry clients). This "scientist" was joined by a confrère from the Heartland Institute, the Exxon-funded group that had put up the billboards comparing climate scientists to

Charles Manson. Year after year they kept up the pressure, with endless radio ads promoting the (false) notion that wind power was raising electric rates for Kansans. Finally, they won, and the mandatory targets for renewable energy were replaced with a "voluntary commitment."[26]

This kind of effort was replicated, successfully, in one state capital after another. In Wisconsin, Governor Scott Walker, a Koch acolyte, ordered employees charged with oversight of state-owned land from even discussing climate change on the job.[27] In North Carolina, a dime-store magnate named Art Pope, a member of the Koch donor network, became the most powerful player in the state's politics, making immense campaign donations. One of his organization's first goals was repealing the state's commitment to modest levels of wind and solar power. It helped sponsor the "Hot Air Tour," set up by the Kochs' Americans for Prosperity, which brought prominent climate deniers to the state. Eventually, North Carolina decided that it would ban state policy makers from using scientific estimates of sea level rise in the coastal planning process.

While this was going on, the network was pushing at least as hard in Washington. The Kansas congressman Mike Pompeo, who had taken more money from the Kochs than any other member of the House (and who is now secretary of state), sponsored legislation upon his arrival in Washington to kill tax credits for wind power, saying it should "compete on its own," which is particularly unfunny given the vast federal subsidies for fossil fuel. Other officials, such as Scott Pruitt, then the attorney general of Oklahoma, were launching one lawsuit after another against the Environmental Protection Agency—in one case, Pruitt literally took a letter from Devon Energy, one of the state's leading frackers, and simply copied it onto his official stationery before mailing it off to Washington.

The point is, long before Donald Trump took over, the Kochs' network had shut off the possibility of any serious action on climate change. I remember, way back in 2004, interviewing Senator John McCain, who had decided that global warming was a crucial challenge. "I do believe that Americans, and we who are policy makers in all branches of government, should be concerned about mounting evidence that indicates that something is happening," he said, and he convened hearings on climate science that led to a modest bill. It lost 55–43, but it seemed like a start. "The race is on," he told me. "Are we going to have significant climate change and all its consequences, or are we going to try to do something early on? Right now, I don't think we're going to act soon enough without significant degradation of our environment. I hope I'm wrong."[28] He was, of course, not wrong—in fact, after a Koch-backed Tea Party challenger came after *him*, McCain himself started blasting away at those he'd once agreed with. When Secretary of State John Kerry called climate change a "weapon of mass destruction," McCain responded, "On what planet does he reside?"[29] By 2014, when McCain made that sneering jape, only 8 of 278 Republicans in Congress were still willing to acknowledge that man-made climate change was real, much less do anything about it.

The campaign against renewable energy was particularly effective given how fast the cost of solar panels and wind power was falling. By 2017, as countries around the world accelerated their efforts to put up windmills and solar arrays, growth in rooftop solar came to a "shuddering halt" in the United States, mostly because of a "concerted and well-funded lobbying effort by traditional utilities, which have been working in state capitals across the country to reverse incentives," according to the *New York Times*.[30] The utilities had turned to the (Koch-funded) American Legislative Exchange Council for model legislation, and in one place after another, Koch-funded utility commissions were putting the brakes on renewables.

Take Arizona, which should be about the easiest place on the planet to make solar power work—Phoenix boasts 299 sunny days a year. On an agreeably cool March morning a few years ago, I stood on the roof of a suburban ranch house in Surprise, a suburb of Phoenix, with Elon Musk's cousin Lyndon Rive, who was at the time the CEO of Solar City, the biggest installer of rooftop solar in the country. Around us, a five-man crew was laying out a grid of solar panels, following a plan designed by an employee in California who had measured the roof by looking it up on Google Earth. The crew had assembled at the house at seven that morning, and by five in the afternoon the new solar array would be ready to be turned on. The homeowner was paying nothing up front, and within the first month, she would see her total electric bill decline—why would anyone not do it? "It's like email in 1991," Rive said. "When I look out at this street, there's no reason every one of these houses can't have solar in a decade." That year, 2015, his company was finishing a solar array somewhere in its eighteen-state service area every three minutes. "That sounds impressive, but it's only two hundred thousand homes so far, out of forty million. My goal is to get it to one home every three seconds. Or maybe we could go faster than that—one every second," he said, snapping his fingers. He pulled an iPhone out of his pocket, called up the calculator app, and punched in some numbers. "At that rate, we could do every house in . . . seventy-six years. No, that's too long—I forgot a division. In a year and a half."

Numbers like that terrified the utilities and the Koch network, and so they set to work. An industry trade group warned that utilities faced "a death spiral." As customers began to generate more of their own electricity from solar panels on their roofs, utility revenues would begin to decline, and the remaining customers would have to pay more for the poles and wires that keep the grid alive, increasing the incentive for the remaining customers to leave. Instead of figuring out (like some California and New York utilities) how to profit from that transition by brokering energy efficiency, Arizona utilities mustered their political power to simply block change. The state's biggest utility, Arizona Public Service (APS), started making big campaign contributions to sympathetic candidates for the Corporation Commission, its regulator. That is to say, it was paying to pick its own bosses. (APS even helped fund the campaign of a candidate for Arizona secretary of state because the candidate's father was a key vote on the Corporation Commission.)

Those regulators in turn soon allowed some of the state's utilities to start levying a hefty fee on customers who wanted to put panels on their roofs, at which point Solar City's business dried up. It started laying off workers and closing down its distribution centers.

That's in Phoenix—in the Valley of the *Sun*, where the basketball team is the *Suns*, where the college basketball and football teams are the *Sun* Devils—which is going to fall into a "death spiral" of its own in short order unless we get climate change under control. Multiply that story by twenty or thirty other states, and you'll understand how solar installations came to that "shuddering halt," even before President Trump put a tariff on solar panels. You'll understand how solar jobs actually fell in the United States in 2017, for the first time since the industry's great expansion had begun.

Similar examples are everywhere. Scott Pruitt, the dutiful Koch errand boy who was the first head of Trump's EPA, killed off plans to raise automobile mileage in an effort to undercut the market for electric cars. At the Department of the Interior, Ryan Zinke asked the staff at a Koch-funded institute to draw up language for his resolution shrinking the size of national monuments to allow for more oil and gas drilling. At the Department of Energy, Rick Perry (who once skipped his own arraignment on two felony charges to attend a Koch event) issued new analyses showing that the United States wouldn't reduce its carbon footprint until 2050, meaning that America would "almost singlehandedly exhaust the planet's carbon budget."[31] The key positions in the Department of Energy's renewable energy office were filled by people coming directly from "the Koch Brothers' numerous anti-clean-energy efforts." The new head of the Office of Energy Efficiency and Renewable Energy, for instance, had gained experience for his post by urging Americans to stockpile incandescent lightbulbs from Amazon on the grounds that an intrusive federal government would next be banning night baseball to save electricity.[32] If you tried to figure out the worst way to respond to climate change, all this is what it would look like.

Actually, the *worst* possible plan would also include trying to squash action in every other country, too. And that's what the entire government-hating network managed to achieve in 2017, when President Trump pulled the United States out of the Paris climate accords. It was as shameful a moment as any in our recent history: the country that produces more carbon than any other announcing that it was now the only country on earth not willing to make even a modest international commitment to solving climate change. As the *Washington Post* made clear in a lengthy special report, getting to that moment required a "two-decade crusade" by precisely the interbred group of antigovernment fossil fuel zealots we've already met. The chair of the "Cooler Heads Coalition," the economist Myron Ebell, who was standing by the president when he made his Rose Garden announcement, worked at the Competitive Enterprise Institute, the place where a staffer had once explained that it would be smartest to "neglect" climate change rather than give government the authority to deal with it. Ebell had formerly worked at a group called Frontiers of Freedom, where he helped run a "complex influence campaign" in support of the tobacco industry. In fact, Philip Morris USA was one of the early funders of the Cooler Heads Coalition, along with companies such as Chevron. They were joined by the Heartland Institute and Americans for Prosperity. (Exxon chimed in with grants to CEI to help support the work.) They worked for years, from Kyoto to Copenhagen to Paris, and they never gave up the fight. When Trump was elected, they drafted a letter reminding him of his campaign pledge to pull America out of the Paris climate accords, despite the objections of most of his advisors: "Mr. President, don't listen to the swamp. Keep your promise." And he did.

The Paris Agreement had been essentially voluntary anyway—the rest of the world had given up on negotiating a real treaty because they saw, after the Kyoto talks in the 1990s, that the power of the fossil fuel industry meant that the U.S. Senate would never muster the two-thirds vote needed to ratify a treaty. Hence, international diplomats knew the best they could hope for at Paris was a set of pledges, and even those were nowhere near rigorous enough to meet the targets they set of holding the planet's temperature increase below two degrees Celsius. Their hope coming out of Paris was that merely starting to work would set in motion a virtuous cycle: as countries saw that renewable energy was cheap and effective, they'd ramp up their commitments. But Trump's pullout slowed that momentum considerably.

It's not that we'll never have a world that runs on sun and wind. We will—free energy is hard to beat, and seventy-five years from now that's what we'll use—but if we tarry along the way, that wind and sun will be powering a badly broken planet. These men happened to be in a place where they could use their power to slow us down precisely at the moment when we needed to speed up, at the moment when one more burst of carbon would break the planet. And so, they've become permanently powerful. Millennia after they've lost the ideological fights, the sea level will still be rising. They've scrawled their names into geological history, ugly graffiti that scientists will be deciphering millions of years into the future (assuming there are scientists). Many of the same people managed to cripple Obamacare, too, which is a tragedy—it means lots of people will suffer unnecessarily and die. But when eventually our politics escapes their grip, it won't be impossible to build a health care system like those in all the other nations of the world. Climate change is different. Once the Arctic melts, there's no way to freeze it back up again, not in human time. The particular politics of one country for one fifty-year period will have rewritten the geological history of the earth, and crimped the human game. That's what leverage looks like.

12

There's one other spot on Earth with game-changing leverage. And though it's not that far as the drone flies from the Palm Springs resorts where the Kochs gather their cronies each year, it's a very different world.

The tech billionaires who inhabit Silicon Valley aren't at all like the fossil fuel moguls and assorted other magnates who celebrated Trump's rise to power. Instead of aging troglodytes, they're mostly youthful social progressives. Don't look for them on the golf course; they're kite-surfing. No, they're not. They *used to* kite-surf, but now they're hydrofoiling. "It's like flying," Ariel Poler, a start-up investor, told a reporter—he was standing by the winged doors of his Tesla and pulling on body armor and a helmet before heading into the ocean. "The board doesn't touch the water. It's like an airplane wing."[1]

Anyway, these tech masters would laugh, and not politely, at the thought of trying to resurrect an eighteenth-century technology like coal. They're all about the future: Tesla is installing the world's largest rooftop solar array on the top of its Gigafactory, which produces more lithium-ion batteries than any facility on Earth. Google spelled out its corporate logo in mirrors at the giant solar station in the Mojave Desert on the day it announced that it would power every last watt of its global business with renewable energy; it's the world's biggest corporate purchaser of green power.[2]

But there is exactly one human being who bridges that cultural gulf between these different species of plutocrat. *Vanity Fair*, in 2016, declared that Ayn Rand was "perhaps the most influential figure in the tech industry." Steve Wozniak (cofounder of Apple) said that Steve Jobs (deity) considered *Atlas Shrugged* one of his guides in life.[3] Elon Musk (also a deity, and straight out of a Rand novel, with his rockets and hyperloops and wild cars) says Rand "has a fairly extreme set of views, but she has some good points in there."[4] That's as faint as the praise gets. Travis Kalanick, who founded Uber, used the cover of *The Fountainhead* as his Twitter avatar. Peter Thiel, a cofounder of PayPal and an early investor in Facebook, once launched a mission to develop a floating city, a "sea-stead" that would be a politically autonomous city-state where national governments would have no sway.[5]

Some of Silicon Valley's antigovernment sentiment is old, or at least as old as anything can be in Silicon Valley. As early as 2001—before the iPhone and Facebook, back in the days when you just checked email—a writer named Paulina Borsook published *Cyberselfish*, a book she called a "critical romp through the terribly libertarian culture of high-tech." Even then, she said, it was unsurprising to open the local newspaper—this was right before Craigslist decimated local newspapers—and see a personal ad that read, "Ayn Rand enthusiast is seeking libertarian-oriented female for great conversation and romance. I am a very bright and attractive high-tech entrepreneur." Every industry has a flavor, and tech's was the hatred of regulation, a "pervasive weltanschauung" that "manifests itself in everything from a rebel-outsider posture" to "an embarrassing lack of philanthropy."[6] Suspicion of government, she said, was "the techie equivalent to the Judeo-Christian heritage of the West. Just as, if you live in the West, you are shaped by this Judeo-Christian heritage regardless of how you were brought up," so Randian hubris flowed through the water in Cupertino and Menlo Park.[7] Borsook credited it to many things: for one, annoyance at the government's clueless early attempts to regulate tech by, say, banning strong cryptographic protection. And then there was the simple fact that coders live, by necessity, in a logical, rule-based universe that "can put you in a continual state of exasperation verging on rage at how messy and imperfect humans and their societies are."[8] It's all a little silly, as it was government investment that got the internet up and running in the first place, but there's no denying that anyone put behind a keyboard for the first time comes away with a sense of autonomy: You can explore anywhere you want to go. It feels *free.*

In any event, the leaders of this community are deeply attached to the idea that they should be left alone to do their thing: create value, build apps, change the world. For them, the key Rand quote is not about the immorality of community—most new tech is theoretically focused on *building* community, after all—or even about the horror of taxes. Instead, it's from early in *The Fountainhead*, when Howard Roark is explaining to his architecture professors that he's going to design buildings the way he wants to. The school's dean, who has accused him of going "contrary to every principle we have tried to teach you, contrary to all established precedents and traditions of Art," then asks, "Do you mean to tell me that you're thinking seriously of building *that way*, when and *if* you are an architect?"

"Yes."

"My dear fellow, who will let you?"

"That's not the point. The point is, who will stop me?"[9]

For reasons that will soon become clear, that may turn out to be the crucial question of the human future.

PART THREE

The Name of the Game

13

I was talking to this guy I know named Ray, and he asked me what I'd been up to that day. I said I'd been out cross-country skiing with the dog.

"Cross-country skiing is fine," he said. "But I don't like downhill. I also don't like being on the side of cliffs. I don't drive anymore on roads that go around the side of mountains. I avoid that, because we don't have backups yet for our version-one biological bodies."

How was he feeling? I asked.

"So far, so good," he said. "I've fine-tuned my regimen. I've gotten it down to about a hundred pills a day. It used to be more."

"A hundred pills?"

"A good example is metformin. It appears to kill cancer cells when they try to reproduce. . . . Nominally it's for diabetes. I've been saying for twenty-five years it's a calorie-restriction mimetic."

"Uh-huh," I said.

"The reason that people who are taking it don't have *zero* cancer cells is that they don't take it quite right. They take a big dose in the morning. You need to take a five-hundred-milligram extended-release pill every four hours. It's more than the maximum dose, nominally."

So, this guy Ray, Ray Kurzweil, is the "director of engineering" at Google, which is arguably the most important company on the planet.

He leads a team charged with developing artificial intelligence. And the reason he is so careful in his daily life is that he firmly believes that if he can just live to 2030 or so, he will never die, that we're accelerating with such great speed toward technological power so immense that it will reshape everything about us. Again, he's not a crank—or, if he is, he's a crank who's directing engineering efforts at the company with the largest market capitalization ever recorded.

"In 1955, when I was seven, I recall my grandfather describing a trip to Europe," Kurzweil told me one day.[1] "He was given the opportunity to handle Leonardo da Vinci's notebooks. He described this experience in reverential terms. These were not documents written by God, but by a human. This was the religion I grew up with: the power of human ideas to change the world. And the notion that you, Ray, could find those ideas. To this day, I continue to believe in this basic philosophy. Whether it's relationship difficulties or great social and political questions, there is an idea that will allow us to prevail."

Of all Kurzweil's many ideas, acceleration is his most profound, "a key basis for my futurism," he says. Essentially: our machines are getting smarter, and they're getting smarter faster. "The number of calculations per second, per constant dollar, has been on a smooth trajectory right back to the 1890 census," he says, a trajectory that he emphasizes is accelerating exponentially, not linearly. His critics, he says, "apply their linear brains. It's like when we were sequencing the genome. People said it would take seven hundred years. But when you finished one percent after seven years, you were almost done; you're only seven doublings from one hundred percent. So, our ability to sequence, understand, and reprogram those genetics is also growing exponentially. That's biotechnology. We're already getting significant progress in things like immunotherapy. We can reprogram your system to consider cancer cells a pathogen and go after it. *It's a trickle now, but it will be a flood over the next decade.*"

Kurzweil's maxim, he insists, applies not just to biotechnology. The basic idea (that the power of a computer keeps doubling and doubling and then doubling again) governs a wide variety of fields, all of which show signs that they're coming into the steep slope of the growth curve. For Kurzweil, it's much like what happened two million years ago, when humans added to their brains the big bundle of cells we call the neocortex. "That was the enabling factor for us to invent language, art, music, tools, technology, science. No other species does these things," he says. But that great leap forward came with intrinsic limits: if our brains had kept expanding, adding neo-neocortexes, our skulls would have grown so large we could never have slid out the birth canal. This time that's not a problem, given that the big new brain is external: "My thesis is we're going to do it again, by the 2030s. We'll have a synthetic neocortex in the cloud. We'll connect our brains to the cloud just the way your smartphone is connected now. We'll become funnier and smarter and able to more effectively express ourselves. We'll create forms of expression we can't imagine today, just as the other primates can't really understand music."

Once again, this is Google's director of engineering speaking. And speaking not just for himself. His boss, Sergey Brin, says the same thing, quite plainly: "You should presume that someday we will be able to make machines that can reason, think, and do things better than we can."[2] To a remarkable extent, we already have. In 2016, the world's best Go player was beaten by a computer program, which went on the next year to beat all sixty of the world's top players, even though Go is supposed to be much harder, subtler, more *human* than chess. In 2017 an artificial intelligence program crushed the world's top players at Texas Hold 'Em—that is to say, it knew how to bluff. Given enough examples, AI programs can now learn almost anything: Facebook's DeepFace algorithm recognizes specific human faces in photos 97 percent of the time, "even when those faces are partly hidden or poorly lit," which is on a par with what people can manage.[3] (Microsoft boasts that *its* software can reliably distinguish between pictures of the two varieties of Welsh corgi.)[4] An AI bot spent two weeks learning a video game called *Defense of the Ancients*, and then defeated the world's top players. "It feels a little like a human but a little like something else," said one of the players who was vanquished.[5]

Sure, it all seems a little trivial—games, after all. The most visible product so far from Kurzweil's team at Google is Smart Reply, those three suggested ripostes at the bottom of your Gmail. ("That sounds great." "Can't make it then." "Let me check!") But Kurzweil's not really out to help you answer your email; he's out to collect more data, to help the cloud learn. *Wired* magazine reported in 2017 that it's "just the first visible part of the group's main project: a system for understanding the meaning of language. Codenamed Kona, the effort is aiming for nothing less than creating software as linguistically fluent as you or me."[6] Sound unrealistic? If so, it won't be for the lack of computing power. Kurzweil has estimated that by 2020, a thousand-dollar PC will have the computing power of a human brain: twenty million billion calculations a second. By 2029, it should be a thousand times more powerful than the human brain, at least by these brute measures. By 2055, "$1,000 worth of computing power will equal the processing power of all the humans on the planet," he says.[7] By 2099, should we get there, "a penny's worth of computing power will be a billion times as powerful as all the human brains now on the planet."

For the moment, let's not try to figure out whether this is a good thing or a bad thing. For now, let's just operate on the assumption that it's a *big* thing, that it represents an unmatched degree of leverage. If the unchecked and accelerating combustion of fossil fuel was powerful enough to fundamentally change *nature*, then the unchecked and accelerating technological power observable in Silicon Valley and its global outposts may well be enough to fundamentally challenge *human nature*. It took a couple of hundred years to do it with coal and gas and oil, though that was an example of acceleration, too—half the emissions, and the ones that seem to have shattered various physical thresholds, came in the last three decades. It probably won't take that long with artificial intelligence, or so the scientists who study the field tell us.

To be clear, we already have achieved what the writer Tim Urban calls artificial narrow intelligence, sometimes referred to as "weak AI." "There's AI that can beat the world chess champion in chess, but that's the only thing it does. Ask it to figure out a better way to store data on a hard drive and it'll look at you blankly," he says.[8] This weak AI is all around us. It's why Amazon knows the thing you want to buy next, and it's how Siri sort of responds to your queries, and it's why your new car knows to slow down if another car pulls in front of you. When the fully self-driving car finally arrives in your driveway, that will be weak AI to the max: thousands of sensors deployed to perform a specific task better than you can do it. You'll be able to drink IPAs for hours at your local tavern, and the self-driving car will take you home—and it may well be able to recommend precisely which IPAs you'd like best. But it won't be able to carry on an interesting discussion about whether this is the best course for your life.

That next step up is artificial general intelligence, sometimes referred to as "strong AI." That's a computer "as smart as a human *across the board*, a machine that can perform any intellectual task a human being can," in Urban's description. This kind of intelligence would require "the ability to reason, plan, solve problems, think abstractly, comprehend complex ideas, learn quickly, and learn from experience."[9] Five years ago a pair of researchers asked hundreds of AI experts at a series of conferences when we'd reach this milestone—more precisely, it asked them to name a "median optimistic year," when there was a 10 percent chance we'd get there; a median realistic year, a 50 percent chance; and a "pessimistic" year, in which there was a 90 percent chance. The optimistic year: 2022. The realistic year (the year when they thought there was a 50 percent likelihood): 2040. The pessimistic year: 2075. That is, the people working in the field were convinced that there was a 90 percent chance we'd have strong artificial intelligence by the time a child born this year was middle-aged (middle-aged by our current reckoning—stay tuned). A similar survey, conducted more recently, simply asked experts when they thought we'd get there. Forty-two percent said 2030 or before; only 2 percent said "never."[10] As one Carnegie Mellon professor put it, "I no longer have the feeling, which I had twenty-five years ago, that there are gaping holes. I know we don't have a good architecture to assemble the ideas, but it's not obvious to me that we are missing components."[11]

What happens then? What happens once a computer is as smart as a person? Probably, say some of these AI experts, it just keeps going. If it's been programmed to keep increasing its intelligence, perhaps it takes it an hour to go from the understanding of an average four-year-old to "pumping out the grand theory of physics that unifies general relativity and quantum mechanics, something no human has been able to definitively do," says Urban. "Ninety minutes after that, the AI has become an artificial super intelligence, 170,000 times more intelligent than a human." As he points out, we have a hard time imagining that, "any more than a bumblebee can wrap its head around Keynesian economics. In our world smart means a 130 and stupid means an 85 IQ—we don't have a word for an IQ of 12,952."[12] You can see how what I've been calling "the human game" might be somewhat altered by such a development, or any development remotely like it. It's leverage on a different scale.

But before we figure out how likely all this is, and before we figure out if it's a good idea, let's look at one particular real-world example of these fast-growing new powers. It will give us a better sense of how far we can go and still stay ourselves.

14

In 1953, Francis Crick and James Watson discovered the double-helix nature of DNA, which was a remarkable achievement, but it didn't change the world overnight. Some highlights on the genetic time line since:

1974: The first genetically modified animal is produced (a mouse).
1996: Some Scottish blokes clone a sheep and name it Dolly.
1999: An artist named Eduardo Kac sticks some jellyfish DNA in a rabbit and makes her glow a phosphorescent green when exposed to black light. "It is a new era and we need a new kind of art," he explains. "It makes no sense to paint as we painted in caves."
Also 1999: Scientists at Princeton, MIT, and Washington University find that they can boost a mouse's memory by changing a single gene—these "Doogie mice," named after a precociously smart TV character now lost to the mists of time, can locate a hidden underwater platform faster than unimproved mice.
2009: Asian scientists produce an even smarter mouse, which they call Hobbie-J, after a character in a Chinese cartoon.

"When these mice were given a choice to take a left or a right turn to get a chocolate reward, Hobbie-J was able to remember the correct path for much longer than the normal mice, but after five minutes he, too, forgot. 'We can never turn it into a mathematician,' the researcher explains. 'They are rats, after all.'"[1]

Genetic work on other organisms was under way simultaneously, of course, and some of it was moving much faster. Monsanto figured out how to make lots of crops resistant to herbicides, which allowed farmers to spray more herbicides, which has boosted the bottom line considerably (Monsanto's, not the farmers'). But for the human organism in particular, there hadn't been a huge amount to show for the genetic revolution; it was slow work because the tools were lacking. As Michael West, the CEO of Advanced Cell Technology, said, "The dream of biologists is to have the sequence of DNA, the programming code of life, and to be able to edit it the way you can a document on a word processor."[2] As you can tell from the archaic use of the term *word processor*, he said this quite a while ago—in 2000, to be exact.

But then CRISPR happened. First, Japanese scientists noticed something odd about some bacteria they were studying: regularly repeating sequences of DNA whose "biological significance is unknown." They called them "clustered regularly interspaced short palindromic repeats," or CRISPR, pronounced like the drawer in your refrigerator where you leave your produce until it wilts. It turned out that they were actually part of the bacteria's immune system. "Whenever the bacteria's enzymes manage to kill off an invading virus, other little enzymes will come along, scoop up the remains of the virus's genetic code, cut it up into little bits, and then store it in those CRISPR spaces." And then the bacteria use the genetic information they have stored like a mug shot, matching up the RNA in any new virus to see if it, too, needs to be chopped up and stored away.[3] Anyway, at a certain point just a few years ago, some scientists recognized that the talent of this enzyme, called Cas9, could be put to good use. If they fed it artificial RNA—a fake mug shot, as it were—it would search for anything with that same code and start cutting.

Which scientist figured it out and exactly when is a matter of some dispute, with billions of dollars riding on the outcome. In 2012, Jennifer Doudna at Berkeley and a Swedish researcher named Emmanuelle Charpentier published a paper showing that they could use the technique to slice any genome at any place they desired. The next year, Feng Zhang, at Boston's Broad Institute, demonstrated that it worked with human and mouse cells; and Harvard's George Church showed a slightly different technique that worked on human cells. What's not in dispute is that CRISPR provides genetics researchers with something resembling that "word processor" they'd always hoped for. "Gene editing went from being laborious and expensive to simple and cheap," *Vox* reported in December 2017. "In the past, it might have cost thousands of dollars and weeks or months of fiddling to alter a gene. Now it might cost just $75 and only take a few hours. And this technique has worked on every organism it's been tried on."[4] As Doudna herself put it, "The genome—an organism's entire DNA content, including all its genes—has become almost as editable as a simple piece of text. . . . Practically overnight, we have found ourselves on the cusp of a new age in genetic engineering and biological mastery."[5]

In the first flush of power, as Doudna describes in her book *A Crack in Creation*, biologists created genetically enhanced beagles with "Schwarzenegger-like supermuscular physiques" by making "single-letter DNA changes to a gene that controls muscle formation." By inactivating a single pig gene, researchers have "created micropigs, swine no bigger than large cats, which can be sold as pets."[6] It doesn't work perfectly yet—stock prices for some genetics firms dropped sharply in the summer of 2018 after researchers found that some "human cells resist gene editing by turning on defenses against cancer, ceasing reproduction, and sometimes dying"[7]—but experts called this setback a bump in the road, and were busily plotting the next advances: a new revolution in crop genetics, for instance, that will raise again the questions of whether genetically modified food is safe to eat (almost certainly yes) and whether it upends traditional agriculture (almost certainly yes). And they're exploring unleashing the power of "gene drives," where scientists can force new traits into wild populations of, say, mosquitoes at "unprecedented speeds, a kind of unstoppable, cascading chain reaction."[8] But we're not going to talk about those, because this particular book is about our species. For *our* game, the real power of CRISPR comes with the ability to change people.

This power comes in two forms, and the distinction between them is key. *The first use of this power is to fix existing humans with existing problems. The second would be to alter future humans.* They are very different, and we will need to think hard about them, because one improves the human game, and the other might well end it.

Let's begin with the first type, the benign one. Scientists refer to it as "somatic genetic engineering," but another name would be "gene therapy." Or you could just call it "repair." In laboratory-grown human cells, CRISPR has already been used to "correct the mutations responsible for cystic fibrosis, sickle cell disease, and some forms of blindness," Doudna reports. "Researchers have corrected the DNA mistakes that cause Duchenne muscular dystrophy by snipping out only the damaged region of the mutated gene, leaving the rest intact."[9] Say someone has sickle-cell anemia. It now seems entirely possible to isolate stem cells from a patient's bone marrow, use CRISPR to repair the cells' mutated genes, and then return the edited cells to the patient, where they will "churn out robust amounts of healthy hemoglobin."[10] This kind of work has just begun to leave the laboratory and enter the real world. In the summer of 2017, the FDA approved the first-ever such treatment, this one designed to modify a patient's own cells to fight leukemia. The drug company Novartis had altered the cells of sixty-three patients, and fifty-two of them went into remission—a legitimate miracle. "We believe that when this treatment is approved it will save thousands of lives around the world," the father of a girl named Emily Whitehead told the FDA panel. When Emily was six, she had very nearly died, but then altered genes left her body cancer-free. "I hope that someday all of you on the advisory committee can tell your families that you were part of the process that ended the use of toxic treatments like chemotherapy and radiation as standard treatments, and turned blood cancers into a treatable disease that most people survive," said Emily's dad.[11]

So, again, let's be clear: this first kind of genetic engineering, the repair of defects in existing human beings, does not present a threat to the human game. Somatic engineering extends traditional medicine, allowing us to cure some diseases we were unable to treat before, or that we could attack only crudely, with massive doses of chemicals or radiation. Yes, there are all the usual complications that come with Big Pharma's profit motives and with our unequal health care system. But this kind of work is going to happen, and it is going to make lives better. Three cheers for Kurzweil's law of accelerating information returns, which made it possible.

Or, maybe, two cheers. Because CRISPR, as I've said, also allows for a second type of power. In this second case, we could *change humans before they are born*, altering their DNA in embryo; in this case, the changes would be passed on forever to their offspring.

The first category, as I've said, is called somatic genetic engineering; this second approach usually travels under the name of "germline" genetic engineering, because the germ line consists of those cells that pass on their traits in the course of reproduction. You could also call it *heritable* genetic modification. "Now, for the first time ever," says Doudna, we possess the power to "*direct the evolution of our own species*. This is unprecedented in the history of life on earth. It is beyond our comprehension."[12]

Ever since Watson and Crick discovered the double helix, ethicists have debated the possibility of designing babies, but it's always been a somewhat remote and academic debate, because no one thought it could actually be done anytime soon. Then, CRISPR. In April 2015, researchers at Sun Yat-Sen University, in Taiwan, announced that they had used the technique to edit the genomes of nonviable human embryos, modifying the gene that produces thalassemia, a blood disorder. In 2017 a team in Oregon repeated the feat, this time focusing on a genetic defect that produces heart disease; their lab was more successful in its technique, with fewer "off-target effects," and the researcher who did the work said he hoped to commercialize the process soon. "I have a very strong opinion on clinical applications. This research was not done to satisfy my curiosity," the Oregon researcher said. "This was done to develop the technology and bring it to clinics. It may take a decade, but we will be there."[13]

In the event, it took considerably less than a decade. In late November 2018, another Chinese researcher, He Jiankui, announced that a newborn pair of twin girls, Lulu and Nana, had been genetically altered in his lab before their birth, making them Earth's first designer babies. The story was bizarre: he'd reprogrammed their genes in an effort to make sure that they wouldn't be able to contract the HIV infection, even though, as the AIDS researcher Anthony Fauci quickly pointed out, "there are so many ways to adequately, efficiently, and definitively protect yourself against HIV that the thought of editing the genes of an embryo to get to an effect that you could easily do in so many other ways in my mind is unethical." Apparently the "fix" only took with one of the newborns; there was speculation that the other might have been damaged in the process.[14] Dr. He had already crossed lines: most government and scientific societies have some form of law or regulation against germline engineering, and the Chinese authorities announced that they were suspending his clinical trial; indeed, there was speculation that he may have been arrested, after a government spokesman called his experiment "extremely abominable."[15]

But clearly the lines are weakening. Doudna said in 2017 that she thought CRISPR shouldn't be used to edit embryos "today, but in the future possibly. That's a big change for me." She had shifted her thinking, she said, after reading letters from people with genetic disease in their family. She'd received one just the other day, from a mother with a son diagnosed with a neurodegenerative disease. "He was this adorable little baby, he was in his little carrier and so cute," she recalled. "I have a son and my heart just broke. . . . And you think, if there were a way to help these people, we should do it. It would be wrong not to."[16]

Which is true—one of the better traits of human beings is our general inability to ignore cute babies in distress. (And all babies are cute.) But it's also true (and this is almost the last technical paragraph) that we already have a way, in widespread use, to prevent genetic disease of precisely this type. It's called preimplantation genetic diagnosis (PGD), and here's how it works: Parents at risk for genetic disease use in vitro fertilization to produce a number of embryos—say, eight of them. A lab grows the embryos for five or six days, to the point where they can be tested to see if they carry the problem genes. The doctor then selects an embryo that's free from the disease and implants it in the mother's womb, and on we go. This has been done millions of times around the world. All the diseases, such as thalassemia, that researchers have shown can be eradicated with germline engineering are already routinely selected against with PGD.

In both cases, the eggs are taken out of the mother and their material manipulated on the laboratory bench; for the mother, the procedures are equally invasive. But PGD is not particularly controversial for a simple reason: You are working with the genetic material provided by the parents. You're not adding something new; you're just eliminating the dangerous possibilities presented by the mathematics of genetics. The only, vanishingly rare, cases where it doesn't work are when both parents suffer from the same recessive genetic disorder. If both parents actually have cystic fibrosis, every single child they conceived would carry the disease as well—there would be no healthy eggs from which to select. But those cases are, indeed, vanishingly rare—these are the people who, absent germline engineering, would have to adopt children or use eggs or sperm from someone else.

PGD works so routinely that journalists routinely ignore it. One study from the Center for Genetics and Society found that 85 percent of articles on human genetic engineering don't even bother to mention that an obvious alternative already exists. In fact, PGD works so well that there are worries it could be misused. Doubtless some people are already selecting the sex of their child, which in a sexist world should worry us. But even those worries pale when compared to genetic engineering, because of the natural limits imposed by the parents' existing genes. PGD allows you six or eight possible people, but they are all within the realm of existing chance.

What makes germline engineering attractive to some *is precisely that it offers the chance to go beyond those limits*, to achieve results that nature acting alone could not produce. Instead of selecting from existing possibilities, it will allow us to add new choices to the menu. Dr. He's alteration that aimed at preventing future HIV infection was the barest start. Let Paul Knoepfler, professor in the Department of Cell Biology at the University of California, Davis, School of Medicine, explain what lies ahead: "In the same way that today you might order a customized pizza with green olives, hold the onions, Italian ham, goat cheese and a particular sauce, when you design and order your future GMO sapiens baby you could ask for very specific 'toppings,'" he says. "In this case, toppings would be your choice of unique traits, selected from a menu: green eyes, hold the diseases, Italian person's gene for lean muscle, fixed lactose intolerance, and a certain blood type."[17]

As we gain a better understanding of how the human genome works, as we get more computer power and understand better the interactions among various genes, the menu will naturally get longer and more startling. Listen, for instance, to Dean Hamer, the former chief of gene structure and regulation at the National Cancer Institute's Laboratory of Biochemistry, describe a scene in the near future, when a young couple—he called them Syd and Kayla—get together to tweak their fetus: "They pondered the choices before them, which ranged from the altruism level of Mother Teresa to the most cutthroat CEO. Typically, Syd was leaning toward sainthood; Kayla argued for an entrepreneur. In the end, they chose a level midway between, hoping for the perfect mix of benevolence and competitive edge." Syd and Kayla were also careful not "to set their child's happiness rheostat too high. They wanted her to be able to feel real emotions. If there was a death[,] they wanted her to mourn the loss. If there was a birth, she should rejoice."[18] As the veteran University of Alabama professor Gregory Pence, a pioneer in the field of bioethics, once put it, "Many people love their retrievers and their sunny dispositions around children and adults. Would it be so terrible to allow parents to at least aim for a certain type, in the same way that great breeders . . . try to match a breed of dog to the needs of a family?"[19]

Pence and Hamer were writing in the late 1990s—I quoted them first in a much earlier book called *Enough*. In those days all this was still speculative: genetic alteration was too difficult for it to be a commercial possibility, and we still had a very limited sense of which genes control mood, intelligence, disposition. Since then, we've learned considerably more, to the point where those early predictions sound a little simple-minded. Now we think more in terms of how genes interact. In the summer of 2018, new studies of the genetics of twenty thousand patients on three continents showed that you could track a person's "polygenic score," measuring information "from across someone's genes to assess their influence on educational success, career advancement, and wealth." Find two kids from the same parents growing up in the same home—"the one with the higher polygenic score tends to go farther," which is to say, gets richer.[20] So, it's not hard to imagine how Big Data and Big Biotech will eventually combine, as Kurzweil insists, to produce a (big) new industry.

There's plenty we still don't know, of course. The day after the news broke about the CRISPR-altered embryos in the Oregon lab, a *New York Times* reporter declared that science was "unlikely" to "genetically predestine a child's Ivy League acceptance letter, front-load a kid with Stephen Colbert's one-liners, or bake Beyoncé's vocal range into a baby," because none of these abilities was located on a single gene.[21] As a Stanford professor explained, we aren't able to examine a stack of embryos and say, "this one looks like a 1550 on the two-part SAT."[22] There are, thankfully, an awful lot of genes involved in making someone smart or sassy.

However, we know a lot more than we used to about which genes regulate, say, the levels of serotonin in our bodies; it's not at all far-fetched to imagine scientists trying to produce some changes in a child's temperament. "This is a pivotal point in the push toward genetically modified humans," said Marcy Darnovsky, the head of the Center for Genetics and Society, the day after the Oregon announcement. "A small group of scientists have taken it upon themselves to move forward with reproductive germline modification technologies. Allowing any form of human germline modification leaves the way open for all kinds—especially when fertility clinics start offering 'genetic upgrades' to those able to afford them."[23]

In fact, given that PGD already lets you deal with disease, CRISPR may well end up being less about saving cute babies from genetic illness and more about "improvement." Jennifer Doudna tells a startling story: Not long after news emerged of her CRISPR breakthrough, one of the PhD students in her lab, Sam Sternberg, got an email "from an entrepreneur I'll call Christina. She wanted to know if Sam would be interested in being part of her new company, which somehow involved CRISPR, and she asked him to meet so she could pitch her business idea." When Sam and Christina sat down "at an upscale Mexican restaurant near campus," she began "speaking passionately over cocktails" about how she hoped her business would offer "some lucky couple the first healthy 'CRISPR baby,'" with "customized DNA mutations, installed via CRISPR, to eliminate any possibility of genetic disease." As she tried to lure him aboard, she stressed that she wanted him to work only on diseases, but he was so rattled that he "excused himself before dessert." He'd perceived a "Promethean glint in her eyes and suspected she had in mind other, bolder genetic enhancements."[24]

The important thing, Doudna stressed, was that CRISPR had in fact opened the door to precisely such enhancements. "Had this conversation occurred just a few years earlier, Sam and I would have dismissed Christina's proposal as pure fantasy," she said. "Sure, genetically modified humans made for great science fiction, but unless the *Homo sapiens* genome suddenly became as easy to manipulate as the genome of a laboratory bacterium like *E. coli*, there was little chance" of it actually happening. "Making the human genome as easily manipulable as that of a bacterium was, after all, *precisely* what CRISPR had accomplished." CRISPR has been used to change the metabolism of monkeys, after all. Given the money at stake, "it seems only a matter of time before humans were added to the growing list of creatures whose genomes" were up for grabs.[25]

I'd hazard a guess that "Christina" will not be the last entrepreneur down this road. In fact, there are already players milling around the starting gate, many of them true heavy hitters from Silicon Valley. The best-known "consumer-facing" genetics company is probably 23andMe, founded by Anne Wojcicki. Anne's father, Stanley, was the chair of Stanford's physics department in the late 1990s; he had a couple of students, Sergey Brin and Larry Page, who would go on to start a thing called Google. In fact, they started it in Anne's sister Susan's garage. (Anne would later marry and divorce Brin; Susan is now the CEO of YouTube, owned of course by Google.) The company 23andMe is best known for its saliva test that unveils your genetics, though one of its patents envisions using this knowledge to help people, in the words of UC Davis's Paul Knoepfler, "select a potential mate from a group of possible mates."

Some of its competitors are pushing the envelope a little further. Take GenePeeks, a genetic research company whose main product, Matchright, examines DNA from you and your potential partner and estimates your chances of producing offspring with genetic disorders. Its cofounder, and chief scientific officer, is a Princeton professor named Lee Silver. If you'd like to know what's in the back of his mind, he laid it out many years ago in a book called *Remaking Eden*. The first germline therapies, he predicted, would be performed to eliminate a few obvious diseases such as cystic fibrosis, and those early and compassionate interventions would cause "fears to subside." (This is apparently what Dr. He intended with Lulu and Nana, though in the short term it seems to have backfired.) Silver envisions what comes next: a mother in a maternity room rejoicing in her new son. "I knew Max would be a boy," she explains to visitors. "And while I was at it I made sure that Max wouldn't turn out to be fat like my brother Tom." A few iterations later and this time a mother is comforting herself during labor by leafing through a photo album of what her infant daughter will look like when she's sixteen: "Five feet five inches tall with a pretty face."[26]

"By the time scientists had employed CRISPR in primate embryos to create the first gene-edited monkeys, I was asking myself how long it would be before some maverick scientist attempted to do the same in humans," Doudna writes. It was time for a "conversation," she felt, and "given that this scientific development affects all of humankind, it seemed imperative to get as many sectors of society as possible involved. What's more, I felt the conversation should begin immediately, before further applications of the technology thwarted any attempts to rein it in."[27] That makes sense to me. Clearly, CRISPR is a perfect example of what Ray Kurzweil meant when he said that exponential increases in computing power would change the world. It's one instance, one of the most striking, of what that new power might produce. It couldn't be more remarkable: a "word processor" for the DNA that is at our core.

So: what could germline engineering do to humans, and to the game we've been playing?

15

The advertisement writes itself: As we get better at germline engineering over the years, we could produce improved children. Their smiles would reveal broad rows of evenly spaced teeth, and of course they'd be smiling a lot because they'd be in a good, sunny mood. And why not, given that their fine-tuned brains would be earning them high grades. "Going for perfection," as James Watson, the father of the genetic age, once put it. "Who wants an ugly baby?" Who indeed. (Of course, you might want to be a little careful here, as someone has to define "ugly." Watson, for instance, also said, "[W]hen you interview fat people, you feel bad, because you know you're not going to hire them," and suggested further that germline engineering could be used to deal with the problem of "cold fish.")[1] We have giant industries based on the idea of what constitutes beauty, and libraries full of self-help books that point us toward particular personalities, so it stands to reason that many people will see this kind of genetic improvement as an obvious next step in our progress as a species.

In the first flush of enthusiasm about new technologies, though, we often overlook the possible drawbacks. For example, if you knew everything you now know about how the smartphone and social media were going to affect your life, and our society, would you still welcome them as enthusiastically as you did the first time you saw an iPhone or logged on to Facebook? That's not a useful question at this point; we have the world we have, Twitter and all. But as we don't yet quite have a world with germline genetic engineering, we should raise the questions now.

It's not as if possible worries are buried very far down. Jennifer Doudna reports that in the years since she pioneered CRISPR, she's had a series of nightmares, most notably one in which Adolf Hitler (with a pig face, "perhaps because I had spent so much time thinking about the humanized pig genome that was being rewritten with CRISPR around this time") summons her to tell him about "the uses and implications of this amazing technology you've developed."[2]

It's never a good sign when even an imagined Adolf Hitler is interested in your work, but for the moment, let's leave aside the specter of cloned soldiers in jackboots and concentrate instead on the more practical problems and immediate difficulties that could arise from human genetic engineering, or from the strong artificial intelligence that scientists say may be just around the corner.

It's worth remembering that any new technology arrives in a world that's already shaped a certain way. If it's a powerful technology, it can either shake up that pattern or help set it in stone. So, for instance, we've seen that most of the planet is at a moment of maximal inequality right now. And we can say with some certainty that engineering your baby will be expensive. Even now, after many decades, IVF treatment for couples with fertility problems runs quickly into the tens of thousands of dollars, usually uncovered by insurance. So, even a pundit with an unimproved IQ can confidently predict that this new technology will make inequality worse. "Since the wealthy would be able to afford the procedure more often," Doudna points out, "and since any beneficial genetic modifications made to an embryo would be transmitted to all of that person's offspring, linkages between class and genetics would ineluctably grow from one generation to the next, no matter how small the disparity in access might be." (She's generously considering how this will play out "in countries with comprehensive health-care systems," which is a polite way of saying "not America.") "If you think our world is unequal now," she adds, "just imagine it stratified along both socio-economic *and* genetic lines."[3]

In truth, this objection is so obvious that the people who plan on carrying out this work don't even bother pretending otherwise. Lee Silver, the Princeton professor who runs GenePeeks, said long ago that eventually "all aspects of the economy, the media, the entertainment industry, and the knowledge industry will be controlled by members of the GenRich class." Meanwhile, "Naturals" will work "as low-paid service providers or laborers." Before too long, he added, the two groups will be genetically distinct enough that they'll have "no ability to cross-breed, and with as much romantic interest in each other as a current human would have for a chimpanzee." Even before mating becomes impossible, he says, "GenRich parents will put intense pressure on their children not to dilute their expensive genetic endowment in this way."[4] The Oxford ethics professor Julian Savulescu, a proponent of human engineering whom we will meet again later, told an interviewer that, "in all likelihood," the technology would exacerbate inequality. His solution: genetically improving the moral impulses of early adopters so that they would "make these technologies available to more people and reduce inequality."[5] This seems a fairly roundabout way of proceeding, though perhaps no odder than the proposal by geneticists of a government-run lottery for a ticket to genetically enhance your kid. (Call it "Charlie and the Baby Factory.")

In fact, if one were genuinely worried about inequality—or indeed, if one were worried about disease in general, or happiness, or children—one wouldn't spend much time and money on biology at all. Genetics plays a part in determining who we are and how our lives proceed, but as Nathaniel Comfort, a professor of the history of biology at Johns Hopkins, points out, "Decent, affordable housing; access to real food, education, and transportation; and reducing exposure to crime and violence are far more important."[6] Consider the experience of the writer Johann Hari, invited to a conference organized by Peter Thiel on depression, anxiety, and addiction. He was amazed to find that most of the participants were convinced that such problems were caused by "malformations of the brain." When it was his turn to speak, Hari said, "As your society becomes more unequal, you are more likely to be depressed." Humans, he continued, "crave connection—to other people, to meaning, to the natural world. So we have begun to live in ways that don't work for us, and it is causing us deep pain."[7] If we wanted to somehow engineer better humans, we'd start by engineering their neighborhoods and schools, not their genes. But, of course, that's not politically plausible in the world we currently inhabit, the world where "there is no such thing as society. There are just individuals." If there are just individuals, that's where you start and end.

The advertisements for ever-greater artificial intelligence write themselves, too: cars that drive you where you want to go, bartenders that mix perfect drinks. As with somatic gene repair for sick patients, there are uses for these new technologies that seem to make perfect sense: the specialized robots that are beginning the decades-long cleanup of the Fukushima reactors, for instance—when one of those robots emerges from the core, it must be "sealed in a steel cask and interred with other radioactive waste,"[8] which you wouldn't want to do with a human. People are building tiny homes with 3-D printers for hurricane refugees; autopilots fly passenger jets most of the time.

Increasingly, though, these technologies are about replacing people who are doing their work perfectly well; it's just that machines can do the work more cheaply. Bricklayers, for instance: a sobering picture on the front page of the *New York Times* recently showed a bricklayer desperately racing like John Henry to match a $400,000 machine called SAM, for "semi-automated mason."[9] A pair of economists recently predicted that by 2033, there was a 99 percent chance that insurance underwriters would lose their jobs to computer programs. Sports referees faced a 98 percent risk of obsolescence, waiters a 94 percent chance, and so on. (Archaeologists were the safest, "because the job requires highly sophisticated types of pattern recognition, and doesn't produce huge profits.")[10] Other researchers pointed out that the Rust Belt has already been so heavily automated that employment will actually drop less there than in places with big service industries. Number one: Las Vegas, which stands to lose 65 percent of its current jobs in the next two decades.[11] So, if inequality worries you, just wait.

These practical losses come with practical gains, obviously: Driverless cars would make it theoretically possible to have fleets of dispatchable, roving electric vehicles that in turn could reduce traffic by 90 percent, free up the city streets that no longer require parking spaces, and save some of the lives lost each year in auto accidents. Also, you could go to a bar and have an extra beer without worry. Still, the transition will be remarkably wrenching. If you include part-timers, more Americans work as drivers than are employed in manufacturing jobs—in forty of the fifty U.S. states, "truck driver" is the single most common occupation.[12] What are they going to do instead? Not become bakers—89 percent of them are expected to lose their jobs to automation by 2033, along with 83 percent of sailors. Wall Street is steadily shedding jobs because algorithms now execute 70 percent of equity trades; it's great for those who remain, given that there's ever more money to go in fewer pockets, but it does make you wonder if we might not be in the last era of high employment.

Tyler Cowen, described by *BusinessWeek* as "America's hottest economist" and proprietor of the country's most widely read economics blog, works in the same Koch-funded economics department at George Mason University where James Buchanan was once a star. His advice to young people is to develop a skill that can't be automated, and that can be sold to the remaining high earners: be a maid, a personal trainer, a private tutor, a classy sex worker. "At some point it is hard to sell more physical stuff to high earners, yet there is usually just a bit more room to make them feel better. Better about the world. Better about themselves.

Better about what they have achieved," he counsels.[13] The author Curtis White, in his book on robotics, concluded: "What survives of the middle class in the future will be a servant class. A class of motivators. A class of sycophants, whose jobs will depend not only on their skills but on their ability to flatter and provide pleasure for elites."[14] Kai-Fu Lee, the head of Sinovation Ventures, an AI venture capital firm, had a slightly sweeter take: "The solution to the problem of mass unemployment will involve 'service jobs of love.' These are jobs that AI cannot do, and that society needs and that give people a sense of purpose. Examples include accompanying an older person to visit the doctor, mentoring at an orphanage, and serving as a sponsor at Alcoholics Anonymous."[15] Laying aside the question of just what it is you're mentoring the orphans about—mentoring them, perhaps, to become orphan mentors in turn—one practical problem is that these don't sound like very well-paid occupations. Lee suggests that high tax rates on the people running AI companies might suffice to make up the difference, although, as he points out, "most of the money being made from artificial intelligence will go to the United States and China," so orphan mentors in the other 190 countries may be out of luck.

Not everyone thinks this will be a problem.

"People say everyone will be out of work. No. People will invent new jobs," Ray Kurzweil told me.

"What will they be?"

"Oh, I don't know. We haven't invented them yet."

Which is fair enough, and in truth, it's as far as we're likely to get with this discussion. This new technology will likely make inequality worse—perhaps engrave it in silicon and DNA. That's worth knowing, but it doesn't answer the question of whether we should proceed. To figure that out, we need to think through other, even deeper, practical problems that come with change at this scale and at this speed. For instance, the end of the world.

Long ago—way back in 2000—Bill Joy, then chief scientist at Sun Microsystems, wrote a remarkable essay for *Wired* magazine called "The Future Does Not Need Us." Joy, the father of the UNIX operating system, argued that the new technologies starting to emerge might go very badly wrong: fatal plagues from genetically engineered life forms, for instance, or robots that would take over and push us aside. His conclusion: "Something like extinction."[16] This was not enough to slow down the development of these new technologies—just the opposite: Joy was writing before CRISPR and back when human beings were still the best chess players on the planet—but it did establish a pattern. Some of the people who know the most about where we're headed are the wariest and the most outspoken. In October 2018, for instance, Stephen Hawking's posthumous set of "last predictions" was published—his greatest fear was a "new species" of genetically engineered "superhumans" who would wipe out the rest of humanity.[17]

Or consider tech entrepreneur Elon Musk, who described the development of artificial intelligence as "summoning the demon." "We need to be super careful with AI," he recently tweeted. "Potentially more dangerous than nukes." Musk was an early investor in DeepMind, a British AI company acquired by Google in 2014. He'd put up the money, he said, precisely so he could keep an eye on the development of artificial intelligence. (Probably a good idea, given that one of the founders of the company once remarked, "I think human extinction will probably occur, and technology will likely play a part in this.")[18] "I have exposure to the very cutting-edge AI, and I think people should be really concerned about it," Musk told the National Governors Association in the summer of 2017. "I keep sounding the alarm bell," he continued, "but until people see robots going down the street killing people, they don't know how to react, it seems so ethereal."[19] All the big brains were talking the same way. Hawking wrote that success in AI would be "the biggest event in human history," but it might "also be the last, unless we learn to avoid the risks."[20] And here's Michael Vassar, president of the Machine Intelligence Research Institute: "I definitely think people should try to develop Artificial General Intelligence with all due care. In this case all due care means much more scrupulous caution than would be necessary for dealing with Ebola or plutonium."[21]

Why are people so scared? Let the Swedish philosopher Nick Bostrom explain. He's hardly a Luddite. Indeed, he gave a speech in 1999 to a California convention of "transhumanists" that may mark the rhetorical high water of the entire techno-utopian movement. Thanks to ever-increasing computer power and ever-shinier biotech, he predicted then, we would soon have "values that will strike us as being of a far higher order than those we can realize as unenhanced biological humans," not to mention "love that is stronger, purer, and more secure than any human has yet harbored," not to mention "orgasms . . . whose blissfulness vastly exceeds what any human has yet experienced."[22] But fifteen years later, ensconced in Oxford as nothing less than the director of the Future of Humanity Institute, he'd begun to worry a great deal: "In fairy tales you have genies who grant wishes," he told a reporter for *The New Yorker.* "Almost universally the moral of those is that if you are not extremely careful what you wish for, then what seems like it should be a great blessing turns out to be a curse." The problem, he and many others say, is that if you have an intelligence greater than our own, it could develop "instrumental goals."[23]

In his foundational 2008 paper, "The Basic AI Drives," researcher Stephen M. Omohundro pointed out that even an AI pointed in the most trivial direction might cause real problems. "Surely no harm could come from building a chess-playing robot," Omohundro begins—except that, unless it's very carefully programmed, "it will try to break into other machines and make copies of itself, and will try to acquire resources without regard for anyone else's safety. These potentially harmful behaviors will occur not because they were programmed in at the start, but because of the intrinsic nature of goal-driven systems." It's really, really smart, and it keeps at its task, which is to play chess at any cost. "So, you build a chess-playing robot thinking you can turn it off should something go wrong. But to your surprise, you find that it strenuously resists your efforts to turn it off."[24]

Consider what's become the canonical formulation of the problem, an artificial intelligence that is assigned the task of manufacturing paper clips in a 3-D printer. (Why paper clips in an increasingly paperless world? It doesn't matter.) At first, says another Oxford scientist, Anders Sandberg, nothing seems to happen, because the AI is simply searching the internet. It "zooms through various possibilities. It notices that smarter systems generally can make more paper-clips, so making itself smarter will likely increase the number of paper-clips that will eventually be made. It does so. It considers how it can make paper-clips using the 3D printer, estimating the number of possible paper-clips. It notes that if it could get more raw materials it could make more paper-clips. It hence figures out a plan to manufacture devices that will make it much smarter, prevent interference with its plan, and will turn all of Earth (and later the universe) into paper-clips. It does so."[25] Those who have seen the film *The Sorcerer's Apprentice* will grasp the basic nature of the problem, examples of which can themselves be almost endlessly (and wittily) multiplied. "Let's say you create a self-improving AI to pick strawberries," Elon Musk once said. "It gets better and better at picking strawberries and picks more and more and it is self-improving, so all it really wants to do is pick strawberries. So then it would have all the world be strawberry fields. Strawberry fields forever."[26]

Remember, in the vision of all these people, computers in the next few years will have brainpower far surpassing that of any person, or any group of persons, and these machines will keep teaching themselves to get smarter, 24/7. As intelligence explodes, and the AI gains the ability to improve itself, it will soon outstrip our ability to control it. "It is hard to overestimate what it will be able to do, and impossible to know what it will think," James Barrat writes in a book with the telling title *Our Final Invention*. "It does not have to hate us before choosing to use our molecules for a purpose other than keeping us alive." As he points out, we don't particularly hate field mice, but every hour of every day we plow under millions of their dens to make sure we have supper.[27] This isn't like, say, Y2K, where grizzled old programmers could emerge out of their retirement communities to save the day with some code. "If I tried to pull the plug on it, it's smart enough that it's figured out a way of stopping me," Anders Sandberg said of his paper clip AI. "Because if I pull the plug, there will be fewer paper clips in the world and that's bad."[28]

You'll be pleased to know that not everyone is worried. Steven Pinker ridicules fears of "digital apocalypse," insisting that "like any other technology," artificial intelligence is "tested before it is implemented and constantly tweaked for safety and efficacy."[29] The always lucid virtual reality pioneer Jaron Lanier is dubious about the danger, too, but for precisely the opposite reason. AI, he says, is "a story we computer scientists made up to help us get funding once upon a time."[30] Imperfect software, Lanier says, not ever-faster hardware, puts an effective limit on our danger. "Software is brittle," he says. "If every little thing isn't perfect, it breaks."[31] For his part, Mark Zuckerberg has described Musk's worries as "hysterical," and indeed, a few weeks after the Tesla baron made public his fears, the Facebook baron announced that he was building a helpful AI to run his house. It would recognize his friends and let them in. It would monitor the nursery. It would make toast. Unlike Musk, Zuckerberg perkily explained, he chose "hope over fear."[32]

A few months later, though, it emerged that Facebook's AI-based ad system had become so automated that it was happily (and automatically) offering mailing lists to people who said they wanted to reach "Jew-haters." Facebook's reliance on automation "has to do with Facebook's scale," one analyst explained. With a staff of 17,000, the company has but one employee for every 77,000 users, meaning it has "to run itself in part through a kind of ad hoc artificial intelligence: a collection of automated user and customer interfaces that shift and blend to meet Facebooker preference and advertiser demand."[33] This is why Zuckerberg is one of the richest men on earth, but it is also a little scary, one example being the Trump presidency. Another came in 2017, when Facebook had to shut down an artificial intelligence system it had built to negotiate with other AI agents: The system had "diverged from its training in English to develop its own language." At first the new lingo seemed "nonsensical gibberish," but when researchers analyzed the exchanges between two bots named Bob and Alice, they determined that, in fact, the bots had developed a highly efficient jargon for bartering, even if it was essentially incomprehensible to humans. "Modern AIs operate on a 'reward' principle where they expect following a course of action to give them a 'benefit,'" one researcher explained. "In this instance there was no reward to continuing to use English, so they built a more efficient solution instead."[34] As Zuckerberg meekly explained when he was summoned to testify before Congress in 2018, "Right now a lot of our AI systems make decisions in ways that people don't really understand."[35] It's not just Facebook. In 2016, Microsoft had to shut down an AI chatbot it had named Tay after just a single day because Twitter users, who were supposed to make her smarter "through casual and playful conversation," had instead turned her into a misogynistic racist. "Bush did 9/11, and Hitler would have done a better job than the monkey we have now," Tay was soon happily tweeting. "Donald Trump is the only hope we've got."[36]

Scientists have even theorized that AIs following their own impulses might explain why we haven't found other civilizations out in space. Forget asteroids and supervolcanoes, says Bostrom—"even if they destroyed a significant number of civilizations we would expect some to get lucky and escape disaster." But what if there is some technology "that (a) virtually all sufficiently advanced civilizations eventually discover and (b) its discovery leads almost universally to existential disaster"?[37] That is to say, perhaps the reason we don't hear from other civilizations is because interstellar space is dotted not with sentient life but with orbiting piles of paper clips.

No one, I think, has a particularly good answer for this set of practical challenges. Unlike global warming or germline engineering, they're not even exactly real—not yet. They're hard to imagine because we've never had to imagine them. Even the engineers building these technologies just work on their particular pieces without ever putting the whole puzzle together. But quite a few of the people who have made it their business to think about these possibilities are scared—of massive inequality written into our genes, of chess-mad AIs. It should be enough to scare us into slowing down, instead of forever speeding up. We should be searching diligently for workable regulations, not cursing government for getting in the way.

Still, I don't want to pursue this line of thinking any further. Practical problems are by definition theoretically soluble—that's why we call them "problems." For the moment, let's assume that we won't create Frankenstein's monsters and that we will make sure that all people have equal access to the fertility lab. Let's assume that, to the degree AI is real, careful programmers will manage to make it into a benign and helpful force that reliably does our bidding. Let's assume that everything goes absolutely right. Let's assume the ads come true.

And then let's ask a more metaphysical, and maybe more important, question: What does *that* do to the human game? What it does, I think, is begin to rob it of meaning.

16

This "human game" I've been describing differs from most games we play in that there's no obvious end. If you're a biologist, you might contend that the goal is to ensure the widest possible spread of your genes; if you're a theologian, the target might be heaven. Economists believe we keep score via what they call "maximizing utility"; poets and jazz musicians fix on the sublime. I've said before that I think there are better and worse ways to play this game—it's most stylish and satisfying when more people find ways to live with more dignity—but I think the game's only real goal is to continue itself. It's the game that never ends, which is why its meaning is elusive.

Still, let's think about those other, more obvious, games: tennis, baseball, stock car racing. They divert a preposterous amount of our time and energy, both physical and mental. All feature some way of keeping score, some way of knowing who's won: most points, most runs, fastest times. They have prizes, championships. But even with all that, their meaning is a little elusive, too. Once the final game of the season has receded a few days into the past, even the most die-hard fan doesn't really care that her team won. (After all, it's only a few months from the end of the World Series to the start of spring training, when the slate is wiped clean and it all begins again.) What we remember are the stories that went into that victory; what lingers on are particular episodes of courage, of sublime skill, of transcendent luck, of great emotion. "It's how you play the game" is the truest of clichés. We assign great meaning to these dramas; they become totems we repeat to one another, and to ourselves, for years. Ask me about the 2004 Red Sox, but not unless you have some time to spare.

For those of us who *play* sports, this is doubly true. The competitions that we train for, sometimes obsessively, need goals: you can't really have a race unless there's a finish line to try to cross ahead of other people. But most people who play sports don't get paid to do it, and no one else is watching; there's no external reward at all. You do it entirely for the meaning, for the exhilarating sense of teamwork that comes from a perfectly executed pick and roll, the lift of the boat when all eight oars are swinging in perfect unison, the sense of discovery that comes from pushing against your own ever-changing limits. I'm a distance athlete—a mediocre, aging one who doesn't race much anymore, but a few times every winter, I'll put on a bib and line up for the start of a cross-country ski race, and an hour or three later I'll cross a finish line somewhere in the middle of the pack. Literally no one cares how well I did, not even my wife. But for me, these are always great dramas, asking the same set of questions: am I willing to make myself hurt, to push past the daily and the easy and the normal? And often the answer is no. I raced last weekend. I was tired, and my mind preoccupied, and half a mile into the race I was twenty yards behind another guy, and there I stayed for the entire race, unable to will myself to go hard enough, hurt enough, to close the gap. No one else could have known or noticed, but I was a little disappointed in myself, just as on other days I've been absurdly if quietly proud. Yes, I'd come in 32nd or 48th or 716th, finishing anonymously in a knot of racers stumbling past some electric eye. But in the race I'd been monitoring in my head, against the guy who came in 33rd or 49th or 717th, I'd managed some great burst of effort, shown myself something I wasn't sure was still there.

So, here's what begins to worry me: with the new technologies we're developing, it's remarkably easy to wash that meaning right out of something even as peripheral as sports. In fact, we're very close to doing it. Erythropoietin, or EPO, is a hormone that stimulates the production of red blood cells. Happily, we have learned to produce it artificially, so we can give it to people suffering from anemia and to those who must undergo chemotherapy. It is remarkable medicine for the repair of problems in our bodies. Apparently, it was given to the cyclist Lance Armstrong when he was being treated for the testicular cancer that almost took his life, and he of course survived, and thank heaven all around. The researchers who figured out what EPO was and how to make it and what dosage made sick people healthy—they were playing the human game with panache.

But if you're healthy and you take EPO, you get extra red blood cells and can run faster and farther than people who don't. Lance Armstrong also took EPO (and testosterone and human growth hormone and probably some other stuff) en route to seven Tour de France victories after his recovery from cancer. It enabled him to climb the Alps with a dash and grit never seen before. People thrilled to watch, transfixed by his epic ascents, and when he launched a charity, Livestrong, they joined by the millions, strapping on yellow plastic bracelets to commemorate the power of the human will. And then it emerged that it wasn't triumph of the human will at all. Sure, he'd worked hard, but he'd done it in concert with those drugs. And for almost all of us, that robbed his victories of any real meaning. He was stripped of his titles, and the charity he'd founded asked him to step aside. "What people connect with is Lance's story," an official of his foundation said. "Take charge of your life." But it turned out that that wasn't really his story; instead, it was "find an unscrupulous doctor who will give you an edge." It wasn't dash and grit; it was EPO. Barry Bonds's home runs were towering, awesome—and then it became clear that they were the product only in part of diligence, application, skill, gift. They were also the product of drugs. We test athletes for those drugs now, in an effort to keep sports "real," to prevent the erosion of their meaning—because otherwise, it is all utterly pointless.

This is not an attempt to be pure, to meet some philosophical ideal. We mix people and machines, for instance, in all kinds of ways. I love Vermont's local stock car track ("Thunder Road, the nation's site of excitement!") because the men and women at the wheel show skill and courage. But I don't think I'd bother going if the races were run by driverless cars. They could doubtless go *faster*, just as runners genetically altered to have more red blood cells can doubtless go faster. But faster isn't really the point. The story is the point.

If something as marginal (though wonderful) as sports can see meaning leach away when we mess with people's bodies or remove them from the picture, perhaps we should think long and hard about more important kinds of meaning. The human game, after all, requires us to be human.

For some people, none of this causes any worry because they perceive no distinction between "artificial" and "natural." Indeed, they say that anything we do is "natural" because we are a product of nature. "The three hundred different breeds of dogs that are around today are all the result of genetic selection over ten thousand years," observes the Oxford ethicist Julian Savulescu. "Some are smart, some are stupid, some are vicious, some are placid, some are hardworking, some are lazy, that's all genetic." He goes on to say, "[W]hat took us ten thousand years in the case of dogs could take us a single generation," once we can engineer human embryos.[1] So, why not?

It's true, obviously, that humans can and do try to engineer their offspring. The mating of two Ivy League grads in the hope of producing a surefire Harvard admit can be as carefully scheduled as the breeding of two Chow Chows to ensure deep-set eyes. Consciously or unconsciously, people reliably try to select mates who will produce the kind of children they want. Indeed, in most of the world's cultures, parents make the matches for their own kids, with the grandchildren very much in mind.

Genetics is not the only tool parents use to try to produce the kids they desire, of course. We also, many of us, invest a great deal of time and energy and money building the right environment. From the moment the embryo has settled in the womb, its genetic code already determined, people start talking to their kids, playing them music. (The smart money is now predicting that "Rosetta Stone language tapes for babies may soon usurp Beethoven as the womb soundtrack of choice.")[2] We try to choose our kids' friends, and their meals, and their pastimes. Some of this is well-intentioned, and some of it is cruel and overbearing—everyone knows people whose lives were stunted by this kind of parenting.

And so, those who want to allow germline engineering often argue by analogy: If it's okay to try to get your kids into Princeton, then surely it's also okay to turn certain genes off or on in order to try to make those kids more intelligent. If we don't limit the ability of parents to push and harass and love their children in a particular direction, why would we limit their ability to accomplish the same thing more efficiently with genetic engineering? It would make Ayn Rand mad as hell to suggest that parents shouldn't be able to do this if they want. Here's James Watson, discoverer of the double helix, who describes himself as a libertarian: "I don't believe we can let the government start dictating the decisions people make about what sort of families they'll have."[3]

But, in fact, both mate selection and parental pressure come with strong limits built in. You can spend a great deal of time looking for the spouse you think will provide your child with the best possible genes, but in the end all you can do is create a set of possibilities, just change the odds some. Nature works within the borders imposed by the genes belonging to the parents; the outcomes are not guaranteed. Even if you're using the PGD technology we discussed, where fertility technicians can help parents create several embryos and then choose the one they like the most, you live within the bounds imposed by your particular genetic codes.

As for nurture, its limits are almost the point, given that people can and do resist their parents' plans for them. For many people, this rejection becomes the turning point in their lives. Rebelling against the wishes and hopes of your parents is how a great many of us define who we are. It may be hard, and it may be painful, and some people may never manage it. And some never need to, because their parents were wise and gentle enough to help them down a congenial path. But it's not impossible.

Whereas, the point of CRISPR, if used for germline engineering of embryos, would be to *replace chance with design*. Because the parents would no longer be playing the odds, and because no child can rebel against a protein.

And if you think about it this way, you soon realize that this is the most antilibertarian technology ever devised. Yes, it increases the ability of parents to make choices. But only by turning the object of their choices, their child, into something we've never seen before: a human built to spec, designed (that is, forced) to be a certain way. Her parents, sitting there in the clinic with their Visa card in hand, will make a series of choices that will then play out over her lifetime and, because those choices will be heritable, over her children's lifetimes, and yea unto the generations. This is control of a kind that tyrants only dream about.

Consider even the early kinds of small changes that would-be baby designers want to target. Though enhanced intelligence is a common goal—"it's not much fun being around dumb people," in the words of James Watson—and though CRISPR advocates say the technique "can in principle be used to boost the expected intelligence of an embryo by a considerable amount," it may actually be hard to get at, as intelligence seems to be spread across a wide array of genes. "Each one accounts for such a small proportion of variance, they are hard to pinpoint," as Steven Pinker explains.[4] Other things are easier: Julian Savulescu describes a variant of the COMT gene associated with altruism, and an MAOA gene variant linked with nonviolence.[5] The gene for the dopamine receptor D_4 (in particular, the "hypervariable coding in its third exon") seems linked straight to mood, as certain variations make people more likely to seek out novelty and to answer yes to statements such as "Sometimes I bubble with happiness" or "I am a cheerful optimist." Other individual genes are also clearly linked to obvious physical traits: MSTN produces "big, lean muscles," Harvard's George Church has noted. When researchers tweak that gene in pigs, they get "double-muscled" swine that "would make body-builders jealous," he said.[6]

And, of course, as our power to transform children accelerates, it may get spookier. Gregory Stock, the former head of UCLA's Program on Medicine, Technology and Society, offered a set of predictions years ago, at the dawn of the genetic manipulation era: "People will be inclined to give their children those skills and traits that align with their own temperaments and lifestyles. An optimist may feel so good about his optimism and energy that he wants more of it for his child. A concert pianist may see music as so integral to life that she wants to give her daughter greater talent than her own. A devout individual may want his child to be even more religious and resistant to temptation."[7] Does this sound absurd? We can put someone in an MRI and see what portions of his brain light up when he prays. In the early summer of 2018, researchers at Columbia and Yale announced they'd discovered the "neurobiological home" for spirituality somewhere in the parietal cortex, directly behind the frontal lobe.[8]

"The best humans have not been produced yet," Michigan State's Stephen Hsu insists flatly. "If you want to produce smart humans, nice humans, honorable humans, caring humans, whatever it is, those are traits that are related to the presence or absence of certain genes and we'll have much finer control over the types of people that are born in the future through this."[9]

Let's assume we can. Even if, at first, we're limited to relatively simple shifts, there's every reason to think that the power will grow swiftly—as Ray Kurzweil points out, it took seven years to decipher the first 1 percent of the human genome, and then just seven years more to finish the job, because the rate of understanding kept doubling. "Everything to do with information technology is doubling every 12 to 15 months, and information technology is encompassing everything," he says[10]—including, of course, our ability to design our kids.

So—and here we reach the crucial point of this whole discussion—*what does it feel like to be that kid*? Let's say it works well. Let's say her parents chose to make her more optimistic, "sunnier." Maybe they were able to add a few IQ points—not a genius yet, but still. And an extra dose of EPO, so her longer, leaner muscles wouldn't tire so easily.

Here's the first thing it feels like: disconnected.

Because time doesn't stop. You get one chance to improve your child, there in the fertility clinic before the egg is implanted, and then she's stuck for the rest of her life with whatever enhancements you've selected. Meanwhile, science marches on. (Fast, fast. In the winter of 2018, a company called Synthego announced that it had figured out how to accelerate CRISPR research so that scientists don't need to "spend weeks" organizing their modifications.)[11] So, by the time your next kid comes along, a year or two later, our ability to manipulate the genome may have doubled. Now you can order up a child with a fancier package of improvements, the human equivalent of a moon roof and leather seats. And so, who is child number one? She's Windows 8, she's iPhone 6, and so on, forever. Her younger brother is smarter, sure, but by the time he's twenty-four and looking for work? The twenty-one-year-olds are going to have an edge, no?

Think about how lonely this feels. On the one hand, you're no longer really related to your past. Current humans have changed so little over the millennia that, say, Stonehenge still makes us feel something. It was created by creatures genetically very much like us, creatures who processed dopamine the same way we do. They are much more like us than our grandchildren would be, should we go down this path. But those modified grandchildren will also no longer be really related to their *future*. They'll be marooned on an island in time, in a way that no human being has ever been before or will be again. When we engineer and design, we turn people into a form of technology, and obsolescence is an utterly predictable feature of every technology we've ever seen. *For a few years, you're more useful than any humans who've ever come before, and then you're more useless.*

But that's just the beginning of the loneliness. With your purchase you will have installed into the nucleus of every cell in your child's body a code that will pump out proteins designed to change her. For a few years that presents no existential problem; she's just chugging along. But then comes adolescence, the moment when we begin to seriously question ourselves, when we try to understand who we are. That's our great task as human beings, and now it can't really be done. She's feeling happy and optimistic? Is that because of some event, some new idea of herself—or is it because she's been constructed to feel that way? How would one know? Every journey of self-discovery would end, ultimately, in the design specs from the fertility clinic. They'd be, in essence if not actuality, the first documents in the baby book and the last testament.

She works hard and takes pride in her achievement—straight As! Why the pride, though, when it's just what she was programmed to do? She takes up running and, holy cow, can she move! Those long, lean muscles never seem to run out of oxygen. But what does that teach her about herself, beyond that she was designed that way? I doubt if Lance Armstrong earned any insights into his character (beyond "I'm a fraud"). In that sense, my athletic career has been far more fruitful than his.

Even the parents seem cheated in this scheme. I take great pride in my daughter Sophie's progress through the world, even though her mother is far more responsible for it than I am, and even though neither of us is all *that* responsible. But we can maintain some sense that our devoted care—all those books read, hikes hiked—helped make her the smart and sunny person she is. Yes, like all of us, she is a creature of her genes, but at least those genes weren't designed to produce a certain outcome. It's one thing to understand that you are who you are in part because of your genes; it's another to understand that you were specifically engineered for a certain outcome. The randomness of our current genetic inheritance allows each of us a certain mental freedom from determinism, but that freedom disappears the day we understand ourselves to be, in essence, a product. Sometimes we need to engineer ourselves: hence Prozac. But you can stop taking Prozac. You can't turn off the engineered dopamine receptor. That's you, and you will never know yourself without it. As climate change has shrunk the effective size of our planet, the creation of designer babies shrinks the effective range of our souls.

And in return we get . . . what? In the best of worlds, where everyone has access to this technology, we'd get more intelligence, more athletic ability. That sounds good. For at least a century, in our high-consumer paradise, we've devoutly believed that *more* is better. For some things, that seems to be true: My phone has more memory; therefore, it's superior. My camera captures more pixels; therefore, hooray! With humans, though, the "therefore" is almost certainly wrong.

I'm assuming that, for many of us, happiness is one goal of our own personal human game. We actually have a fairly good idea of what makes human beings happy, thanks in large part to Mihaly Csikszentmihalyi, the longtime head of the psychology department at the University of Chicago. Back in the 1960s, he was studying painters and noted the "almost trance-like state" they entered when their work was going well. They didn't seem to be motivated by finishing the painting, or by the money they'd get for selling it. It seemed to be the work itself that spurred them on, even in the face of hunger or fatigue.

To follow up on this clue, Csikszentmihalyi and his colleagues developed a method they called "experience sampling." They'd give their study subjects a pager and then buzz them at random intervals throughout the day. When the buzz sounded, they were supposed to quickly fill out a short form listing what they were doing and their mood. Such surveys yielded immense insights—for instance, if people were feeling chaotic and out of control in midafternoon, they were going to spend a lot of the evening watching TV, apparently because it reordered their lives. But the most remarkable finding, robust after many years, was that people were happiest when they were engaged in what Csikszentmihalyi came to call "flow"—that is, when, like those painters, they were fully engaged, and at the limit of their skills. A person in a state of flow has neither less challenge than she can handle, nor more. So, if you're a beginning rock climber, a single boulder can provide you with enough challenge to become fully absorbed; once you master it, you need a steeper wall. Dancers require choreography that they can actually perform; basketball players require opponents good enough to test their skills. It's "a stretching of oneself toward new dimensions of skill and competence," Csikszentmihalyi said.[12]

No one can do this all the time, of course, hence the need for *The Bachelorette* and a bottle of beer. But it's what defines us at our best.

And so, it should therefore sting the Kurzweils of the world to grasp that you can't make a more realized human being by giving him extra talent. The greatest cross-country skier on earth doesn't get more out of a race than I do, even if he finishes it in half the time. As long as I'm fully engaged, the world drops away—and the point is the world dropping away. If you could engineer a rock climber to have stronger fingers and no fear of heights, then she would be able to climb more routes than she could climb now. But so what? She wouldn't get extra satisfaction from her new talent, because the satisfaction comes from being at the edge of her abilities. In fact, you might complicate her life considerably, because she'd have to go farther afield to find cliffs big enough to match her souped-up abilities. If you were eventually able to engineer her to the point where dashing up Mount Everest presented no great challenge, you would have robbed the entire exercise of its point. Flow doesn't increase if you have more ability; it simply requires challenge sufficient to your ability.

We are already capable of being as absorbed and engaged as we ever could be. We're good enough.

•

17

One reason that techno-utopians don't worry about the loss of human meaning is because they're not particularly attached to humans.

There are, to be sure, plenty of doctors hoping for new ways to treat human suffering. But the streak of misanthropy that runs through the conversation of the digital and technological elite is hard to miss: Human brains, the artificial intelligence pioneer Marvin Minsky once explained, are simply "machines that happen to be made out of meat."[1] Robert Haynes, president of the Sixteenth International Congress of Genetics, said in his keynote address that "the ability to manipulate genes should indicate to people the very deep extent to which we are biological machines." It's no longer possible, he insisted, "to live by the idea that there is something special, unique, or even sacred about living organisms."[2] Indeed, in the spring of 2018 a University of Washington professor proposed using CRISPR to create a "humanzee," a human-chimp hybrid, specifically to prove that people aren't special. "The fundamental take-home message of such a creation would be to drive a stake into the heart of the destructive disinformation campaign" holding that people are different from the rest of creation, he explained.[3] This kind of self-loathing permeates the whole subculture. Robert Ettinger, the first man to start freezing his fellow humans so they could be revived in a century or two, looked forward to a golden posthuman age, one where, among other things, we would be reengineered to achieve the "elimination of elimination." He found defecation so unpleasant that he wanted "alternative organs" that would "occasionally expel small, dry compact residues."[4]

By this logic, if we are machines, then our destiny is to be surpassed by better machines. And we shouldn't complain; we should welcome it. The approaching epochal moment when computers will be as smart as humans becomes just a meaningless way station. As the science writer Tim Urban points out, an AI "wouldn't see human-level intelligence as some important milestone—it's only a relevant marker from our point of view—and wouldn't have any reason to stop at our level. And given the advantages over us that even human-intelligence-equivalent artificial general intelligence (AGI) would have, it's pretty obvious that it would only hit human intelligence for a brief instant before racing onwards to the realm of superior-to-human intelligence."[5]

After all, AGI's got better components. Already today's microprocessors run about ten million times the speed of our brains, whose internal communications "are horribly outmatched by a computer's ability to communicate optically at the speed of light," Urban observes. And our human constraints aren't going away: "the brain is locked into its size by the shape of our skulls," while "computers can expand to any physical size, allowing far more hardware to be put to work." Also, humans fatigue easily; also, our software can't be as easily updated. And a group of computers can "take on one goal as a unit, because there wouldn't necessarily be dissenting opinions and motivations and self-interest, like we have within the human population."[6] James Lovelock, the British scientist who formulated the Gaia theory, insisted that robots would inevitably take over simply because it takes a neuron a second to send a message a foot in our brains, while an electron can speed along a foot of wire in a nanosecond. "It's a million times faster, simple as that," he said. "So to a robot, once fully established in that new world, a second is a million seconds. Everything is happening so fast that they have on earth a million times longer to live, to grow up, to evolve than we do."[7]

In other words, forget about the fact that the self-driving truck is going to take away your job. The practical risks we're running pale next to the questions about human meaning: what on earth would be the *point* of people in this new world? The futurist Yuval Harari provides one answer: we could devote our lives to playing ever-more-immersive video games. "If you have a home with a teenage son," he writes, "you can conduct your own experiment. Provide him with a minimum subsidy of Coke and pizza, and then remove all demands for work and all parental supervision. The likely outcome is that he will remain in his room for days, glued to the screen. He won't do any homework or housework, will skip school, skip meals, and even skip showers and sleep. Yet he is unlikely to suffer from boredom or a sense of purposelessness."[8] Steve Wozniak, cofounder of Apple, predicts that robots will graciously take us on as pets so we can "be taken care of all the time."[9] He added that he was now feeding his dog filet mignon, on the principle of "do unto others." None of that is *why* we're developing artificial intelligence. (We're developing it to make money, one business at a time.) But that is *what* many of the people who look closely at it think may happen.

You can already sense the beginnings of this shift. The average person now touches, swipes, or taps his phone 2,617 times a day.[10] Eighty-seven percent of people with smartphones wake up and go to sleep with them. This is by far the largest change in the texture of everyday life during my six decades on earth; nothing else comes close. The artificial intelligences at the other end, the giant algorithms that run Google and Facebook and the like, by now know when we're bored; they understand that we crave the positive reinforcement of "likes"; they know what to feed us to keep us clicking. As Jaron Lanier points out, because the business models of the social media giants prize "engagement" above all, they've learned to shovel negative information at us because "emotions such as fear and anger well up more easily and dwell in us longer than positive ones. . . . Fight-or-flight responses occur in seconds," which is about the right time frame for Twitter, as opposed to, say, a novel or a record album.[11] In the political realm, they've learned that we respond to an ever-greater sense of outrage; hence, Trump.

But Trump is the least of it. The path we've started down is the not-so-gradual replacement of humans with something not so slightly different: a man with a phone more or less permanently affixed to his palm is partway a robot already. Even our posture has begun to change. A 2016 study in the *Journal of Physical Therapy Science* found that there were "significant differences in the craniovertebral angle, scapular index, and peak expiratory flow depending on the duration of smartphone usage."[12] That is, having taken a few million years to stand up straight, we are hunched once more—text neck, iPosture. And we've already subcontracted a good bit of our memory to the Web: seven in ten people can remember phone numbers from their childhood, but not the numbers of their current friends now plugged into their phones. We spend roughly ten hours a day looking at a screen and roughly seventeen minutes a day exercising—that is, using our bodies. "Our lives now are only partly biological, with no clear split between the organic and the technological, the carbon and the silicon," the venerable *National Geographic* intoned in a recent special issue on "the next human." "We may not know yet where we're going, but we've already left where we've been."[13]

And how's that working for us? It's hard to study, both because it's so new and because, aside from the Amish, there's no control group. But the data so far are sobering. The psychology professor Jean Twenge reported that beginning in 2012, when the number of Americans with a smartphone passed the 50 percent mark, there were "abrupt shifts in teen behaviors and emotional states" completely unlike anything that showed up in decades of analysis of generational data. The good news is that teens are physically safer because they're drinking less and having far less sex, and the bad news is that that's because they rarely go out. The number of teens who get together with their friends every day dropped by 40 percent from 2010 to 2015, a curve that's accelerating.

They're in their bedrooms, but not studying, not working. They're, of course, texting and looking at social media, "alone and often distressed." The more Facebook they look at in a day, the unhappier they feel, and this unhappiness isn't just a mild malaise. Eighth-graders who are heavy users of social media increase their risk of depression by 27 percent. Teens who spend three hours a day or more on electronic devices are 35 percent more likely to be at risk of suicide; depressive symptoms among girls spiked by 50 percent. Three times as many teenagers killed themselves in 2015 as in 2007.[14]

None of this is an indictment of young people. Millennials can use their connectedness to do remarkable things, as everyone saw in the months after the Parkland, Florida, school shootings in the winter of 2018. Instead, it's an indication: because they're the first emerging citizens of this particular technological world, they offer us a glimpse into what it will be like. But ask yourself, if you're old enough to remember and your memory still works, what it was like to live before email and Twitter and text? Not for nostalgic reasons, but instead so we can anticipate what it will be like to move into a world ever more dominated by technology.

When we last talked, Ray Kurzweil told me his vision: "As we get smarter, we can create more profound intellectual expressions—music, literature. Beauty and artistic expression of all kinds." Indeed, this is often how technologists imagine the future: Freed from the need to work, we'll paint paintings, play the saxophone, write books all day. Art will be the last human refuge, in the way that Roman nobles composed poems while their slaves raised the grapes and olives for the wine and oil.

But why couldn't computers produce "better" art than people? They can, after all, analyze what we like and then reproduce it. Already there are AIs that compose Bach-like cantatas that fool concert hall audiences, and the auction house Christie's sold its first piece of art created by artificial intelligence in the fall of 2018. But at a deeper level, *that's not even how art works.* The point of art is not "better"; the point is to reflect on the experience of being human—which is precisely the thing that's disappearing.

Even—and here is more rich, sad irony—science is at risk. The profound pleasure that keeps people working on precisely the technology that now threatens to supplant us will vanish, too. You don't think robot biologists will soon replace the real ones?

What are we left with? Nick Bostrom, the early apostle of transhumanism, offers the best case: a superintelligence could "assist us in creating a highly appealing experiential world in which we could live lives devoted to joyful game-playing, relating to each other, experiencing personal growth, and living closer to our ideals." Or we could just smoke weed.

So, why do we do it? Why do we keep on with this ever-accelerating rush into territory that everyone involved understands is risky?

Partly, it's inertia—a body in motion stays that way, and we've been in motion for some centuries now, headed toward something called Progress. It's hard for us to imagine the alternative (though not, as we'll see, impossible).

Partly, it's money, the great magnetic attraction that ensures the motion never ceases. Whatever business you're in, its future success depends on mastering these new technologies. Capitalism played by its current rules doesn't allow anyone to step easily aside.

But there's also something weirder.

18

We know of no story older than *Gilgamesh*, the epic poem of the Sumerians, which dates back about four thousand years. It begins with the friendship of Gilgamesh the king and Enkidu. In the middle of the book, however, Enkidu dies, and after sitting by the body seven days and nights, Gilgamesh sees a worm drop from the nose of his friend's corpse. A great fear rises in him, and he speaks words that could be spoken today:

> *Must I die too? Must Gilgamesh be like that?*
> *It was then I felt the fear of it in my belly. I roam the wilderness because of the fear.*
> *Enkidu, the companion, whom I loved, is dirt, nothing but clay is Enkidu.*
> *Weeping as if I were a woman I roam the paths and shores of unknown places saying:*
> *"Must I die too? Must Gilgamesh be like that?"*

Determined to find the secret to immortality, he undertakes a perilous journey, battling a pride of lions, passing through a tunnel guarded by two scorpion-men, destroying stone giants, and felling 120 trees, which he uses to propel himself across a deadly sea. He survives a storm so horrific that it sends the terrified gods scurrying to the heavens—but all of it is in vain. "Eternal life, which you look for, you will never find," he finally learns from the one man who knows. "For when the gods created man, they let death be his share." Indeed, he's told, the constant hunting for immortality only ruins the joy in life.

From Gilgamesh onward. We are, of course, the animal with consciousness, which is to say the animal that knows that it will die. We don't dwell on it constantly, but it shapes us and the cultures we've constructed. The great psychologist Ernest Becker was convinced that Freud had it wrong: it wasn't sex that our minds repress, but the fear of death, and from that fear we've constructed everything from mighty pyramids to the mightier idea of heaven. The pattern of our lives is set by the span we hope to live: we know how much time we can allot to education, and we can tell the prime of our lives, and if we're brave enough to acknowledge it, we can prepare for our approaching death.

It's true that the average human life span has increased, mostly because far fewer babies die and because advances such as basic sanitation have dramatically reduced disease. Researchers studying chlorination have found that clean water led to a 43 percent reduction in mortality in the average American city, a reminder of what happens when we work together.[1] But the people who live the longest aren't living any *longer.* A hundred and fifteen years appears to be pretty near the upper edge, a boundary set by the so-called Hayflick limit on the number of times human cells can divide; so far it has been as inviolate for humans as the speed of light. And we have, more or less, made our accommodations, as individuals and as societies. The people we admire most are the ones who seem to have come to terms with their mortality—we nod when we hear Martin Luther King Jr.'s words: "A man who does not have something for which he is willing to die is not fit to live." Death is what finally takes the measure of our self-absorption: Ayn Rand and the first libertarians may have wanted to do away with taxes, but even to them mortality seemed a given. David Koch announced his retirement from business and politics in 2018 because of declining health.

Not so, however, the Silicon Valley wave, who are driven in no small part by precisely the fear that haunted Gilgamesh but who have persuaded themselves that they finally have immortality in their sights. Ray Kurzweil takes a hundred pills a day, the better to ward off aging long enough for his peers to figure out how to guarantee he'll never die. That's not particularly unusual behavior among the tech elite—it's easy to find people taking resveratrol, or off-label diabetes medications. As *Wired* magazine reports, "The most daring are rumored to use rapamycin, a powerful drug that prevents organ transplant rejection," even though it suppresses the immune system. The theory is that it "initiates a process where dysfunctional cellular components are degraded or recycled."[2] Peter Thiel, the PayPal billionaire and Trump supporter, either does or does not transfuse himself with the blood of young people, in an effort to retain his youthful vigor. One tech journal said that he was paying forty thousand dollars every three months for the blood of eighteen-year-olds, but he told another reporter that he hadn't "quite quite quite started yet."[3] But, intrigued? For sure. "I'm looking into parabiosis stuff, which I think is really interesting," Thiel said. "This is where they put the young blood into older mice and they found that had a massive rejuvenating effect." A Silicon Valley start-up called Ambrosia has at least one hundred clients who will pay eight grand a pop for the blood of young'uns.[4]

Thiel's Breakout Labs invests in many other start-ups trying to conquer aging—and why not, given that he believes that "probably the most extreme form of inequality is between people who are alive and people who are dead"? (This is saying something when his own net worth exceeds the GDP of roughly thirty countries.) "I stand against confiscatory taxes, totalitarian collectives, and the ideology of the inevitability of the death of every individual," he explained in an essay detailing his reasons for being a libertarian.[5] "I've always had this really strong sense that death was a terrible, terrible thing."

Among Silicon Valley tycoons (again, arguably the most powerful people on Earth), defeating death is high on the to-do list. Yes, they want to make even more money. Yes, they're enchanted by the sheer pleasure of building new technology. And no, they do not want to die. If you want to know why they push ahead so relentlessly with artificial intelligence and germline manipulation, despite the obvious dangers, then you need to listen to them speak. You need to sense precisely how freaked out they are.

The *New Yorker* reporter Tad Friend memorably described an evening at Norman Lear's house, set high in the hills above Los Angeles. The party was a kickoff event for the National Academy of Medicine's Healthy Longevity Grand Challenge, which will award millions of dollars for breakthroughs in the field. There were Hollywood stars in attendance—Goldie Hawn demanded that a Nobel Prize geneticist offer an opinion on glutathione, a powerful antioxidant that features in many health regimens—but the real celebrity was Google cofounder Sergey Brin; you'll recall that his ex runs 23andMe, the pioneering genetics firm. At this gathering, his current girlfriend, Nicole Shanahan, said Brin had phoned her recently with the sad news that he was going to die—someday. Or maybe not, given that Google was investing huge sums in life-extension technologies. In 2009 it hired Bill Maris to run its venture capital fund, and he quickly began devoting most of its vast resources to life sciences start-ups. Why? Because "If you ask me today, is it possible to live to be 500? The answer is yes."[6] "We aren't trying to gain a few yards," Maris says. "We are trying to win the game. And part of it is that it is better to live than to die." Again—this is not some outlier cult. "There are a lot of billionaires in Silicon Valley, but in the end, we are all heading to the same place," Maris says. "If given the choice between making a lot of money or finding a way to make people live longer, what do you choose?"[7]

Doubtless the actual answer is both. Google has gone beyond merely funding other people's start-ups. In 2013 it launched its own venture, Calico, which stands for California Life Company. It's been highly secretive—all anyone really knows about the operation is that it has squadrons of mice eating different diets—but its focus is "the challenge of aging,"[8] and it's definitely not alone. The world's richest man, Jeff Bezos, has diverted some of his cash to the San Francisco–based start-up Unity Biotechnology, which is hard at work on "a cure for aging." At a recent seminar on "the business of longevity," hosted by *The Economist*, an "acolyte" of Peter Thiel (who has also invested in the company) rated it as one of the most likely start-ups to get a drug to market soon. The start-up was, she said, "one of the most exciting companies in a space that's gone from fringe science to hot new field."[9] Typical headline, this one from the British papers: BEATING AGEING IS SET TO BECOME THE BIGGEST BUSINESS IN THE WORLD, SAY TYCOONS.[10]

One way to judge the seriousness of these men and women is to look at their ankles. A great many of them wear a thin leather band with a metal tab engraved with the contact information for Alcor, the world's leading cryonics facility. Much like the flying car, the idea of cryogenics has been around a long time without quite being real. For a 1948 edition of *Startling Stories*, Robert Ettinger (the fellow hoping for the reengineering of human beings so they could defecate small, dry, odorless pellets through an "alternative" orifice) wrote a short story about freezing people, followed up in 1962 by a nonfiction account called *The Prospect of Immortality*. That prospect was slim at first—Alcor technicians had to open the chamber of their initial client and toss in more dry ice to keep him from thawing—and the business has always been plagued by controversy. Who can forget some of the children of Ted Williams suing some of the other children to keep his head from being iced? Things were so sketchy that Timothy Leary, an early Alcor client, gave up and had his ashes shot into space from a rocket (along with those of *Star Trek* creator Gene Roddenberry). His former cryogenic colleagues accused him of succumbing to "deathist ideology."

The industry is maturing, though. Alcor currently has 147 human beings on ice, each of whom paid $200,000 to preserve their whole bodies or $80,000 for the "neuro option," which involves sawing off the head. (There's also a $10,000 discount if you're willing to die in Scottsdale, Arizona, so they can frost you on-site.) "Our view is that when we call someone dead it's a bit of an arbitrary line. In fact they are in need of a rescue," says Alcor's CEO, Max More (a name he bestowed on himself as a reminder of "what my goal is: always to improve, never to be static").[11] Ralph Merkle, a hero in Silicon Valley as one of the inventors of public-key cryptography for computers, is on the board of Alcor, and as a public service, he calculated what it would take to preserve everyone on the planet. Given that fifty-five million humans die annually, it's easier if you just save the noggins; and with a double-walled cooling flask thirty meters in diameter able to accommodate 5.5 million brains, you'd need to build just ten per year to store "the head of every person who died in the entire world, going forward, until such time as their deaths could be remedied." Amortized across the Earth's entire population, Merkle estimates a "surprisingly competitive" price of $24 to $32 per person.[12] Currently, at least a thousand people are waiting for their chance, and they include a large selection of Silicon Valley pioneers.

This being the tech industry, though, a newer iteration of the idea is already available. Nectome is one of the handful of start-ups chosen to be part of Y Combinator, the most important of California's tech incubators. (They're the people who first championed Dropbox, Airbnb, and Reddit.) In fact, Y Combinator head Sam Altman has already plunked down his $10,000 for Nectome's service, which involves embalming your brain when you're near death so that it can later be digitized and encoded. "The idea is that someday in the future scientists will scan your bricked brain and turn it into a computer simulation," writes Antonio Regalado in *MIT Technology Review*.[13]

In fact, this notion that we will one day be meshed with computers and thus live forever has gained currency perhaps because, while bizarre, it seems somehow less absurd than the idea of Ted Williams lumbering around again in the real world. (Presumably alongside Oscar, Max More's goldendoodle, who also has a storage flask awaiting him.)

Ray Kurzweil is an Alcor customer, but it's clearly a fallback position; his real hope is not to die at all and instead to live long enough to reach the point where his failing cells can be repaired by nanobots in the blood. In fact, he says, those nanobot blood cells could perhaps power their own movement, dispensing with the need for a heart, which is after all just a large pump prone to failure. And Kurzweil's pretty sure we'll someday be able to connect our brains directly to the cloud. By the time we can implant a hundred thousand electrodes per square inch of scalp, there will be "no need to read a book—the computer just squirts its contents into your head."[14] Remember, this is the chief scientist at what is by some measures the biggest company in the planet's history.

To talk with Kurzweil is to remind yourself that there is something sweet and wistful to these dreams of immortality. His father, Fredric, died when Ray was young, and the son has filled a storage locker with boxes of his effects (letters, photos, even electric bills), in the hope of someday creating "a virtual avatar of his father and then populating the doppelganger's mind with all this information."[15] So he can talk with him again, father to son. "I do think death is a tragedy," Kurzweil told me. "That's our immediate reaction to it. If someone dies, our immediate reaction—it's considered a tragic thing, not a triumphant thing."

That is of course true sometimes, though not, I think, if someone has lived what seems a full life. You can read the obituary page with a sense of dread, but you can also read it as the chronicle of a world that works.

The obvious practical problems posed by no one dying can be waved away if you believe, like Kurzweil, that in just a little while, artificial intelligence will be providing us with a planet so rich in resources that no one will ever want for anything. "Overcrowded?" he asked me. "Take a train trip anywhere in the world and look out the window. Forty percent of the land we use is for horizontal agriculture. We can do a better job without any of that." By growing our food on vertical stands, he means. "As we extend longevity, we radically expand the resources of life."

That's an ethical improvement on, say, Michael West, currently chair of a California start-up called BioTime, which specializes in "regenerative medicine." West, who organized the first effort to isolate human stem cells for cloning purposes, was once asked whether immortality wouldn't lead to overpopulation. Sure, he said, but "why put the burden on people now living, people enjoying the process of breathing, people loving and being loved. The answer is clearly to limit new entrants to the human race, not to promote the death of those enjoying the gift of life today."[16] That level of selfishness makes Ayn Rand look like Mother Teresa.

And it's that incredible self-absorption that should be the clue to what a bad idea this all really is. I've taken the time to lay out the various advances we may be capable of if we fully embrace the newest technologies—we can "improve" our children; we may be able to live without work (or we may have to); we may be able, in some sense, to live forever—but none of that is living, not in the human sense.

These threats to the human game are existential. Though the technologists at some level value individual humans too much—no one can be allowed to die; we must collect their heads in a giant thermos—they value humanness far too little. They don't understand that some sadness and loss is not just bearable; it's essential. There is an everyday heroism, if you think about it, in bringing up your children fully aware that they will supplant you. That's what human civilization is. If it weren't—if your children were just going to be other beings who perpetually trailed you through infinity by twenty or thirty years—then the most powerful of human connections would in effect be severed. What would you owe them, and vice versa? Those who exalt humans too highly devalue humanity.

A world without death is a world without time, and that in turn is a world without meaning, at least human meaning. Go far enough down this path and the game is up.

PART FOUR

An Outside Chance

19

I don't know—no one knows—if it is still possible to fundamentally alter our trajectory. Climate change is far advanced, and the march of some of these new technologies seems as rapid as it is unregulated. But no one knows that it is impossible, either, and so the last section of this book will be about resistance, about the tools and ideas that might help us keep global warming and technological mania within some limits and, in the process, keep the human game recognizable, even robust.

Resistance is a subject I take up with some reluctance, because I know at least a little bit about its costs. I've spent much of the past thirty years as a volunteer in the fight against global warming. We've had more successes than I imagined we would, some of which I will describe in passing, but we have yet to turn the tide: the power of people is not yet mobilized in sufficient strength to outweigh the financial majesty of the fossil fuel industry, and so we continue down an ever-hotter path. Also, the price of even that mobilization has been enormous: in some parts of the world, environmental advocates are routinely murdered, and even in places where they operate with more freedom, the stress and strain are very real. I know so many people who have given over the prime of their lives to this fight. Some have been to jail, wrecked their careers, burned out their own emotional cores. They've been sued and surveilled by oil companies, attacked by guard dogs. I also know many people who've found their lives in this work, in burgeoning movements that are full of love and friendship. But none of it has been easy—and the climate fight at least has the advantage of being against something clearly ugly and wrong. Opposing wildfire and drought and inundation is conceptually easier than figuring out how to slow down the rush toward a pill that would let you live forever or a genetic tweak that would make sure your child turned out as cute as a button.

Still, even the most powerful foes and effective propaganda have weaknesses that can be exploited. These battles over the human future potentially shake up our usual political categories. For instance, because I am concerned about inequality and about the environment, I am usually classed as a progressive, a liberal. But it seems to me that what I care most about is preserving a world that bears some resemblance to the past—a world with some ice at the top and bottom and the odd coral reef in between, a world where people are connected to the past and future (and to one another) instead of turned into obsolete software. And those seem to me profoundly conservative positions. Meanwhile, oil companies and tech barons strike me as deeply radical, willing to alter the chemical composition of the atmosphere, eager to confer immortality. There is a native conservatism in human beings that resists such efforts, a visceral sense of what's right or dangerous, rash or proper. You needn't understand every nuance of germline engineering or the carbon cycle to understand why monkeying around on this scale might be a bad idea. And indeed, polling suggests that most people instinctively oppose, say, living forever or designing babies, just as they want government action to stabilize the climate.

This political mashup is either a split that will be widened by those who want to own the future, or it's a potential source of great strength. The key, I think, lies in how we see ourselves. If, as the antigovernment rhetoricians insist, we view ourselves only as individuals, then the game is lost: we will never combine in numbers large enough to overcome the deep power and unrelenting focus of great wealth.

But the opposite of libertarian hyperindividualism is not necessarily the Red Army kicking in the door of your father's drugstore. It could also be a sense of social solidarity, an ethic of "We're all in this together." As Pope Francis said, after a closed-door meeting with oil company executives about climate change in 2018, "Decisive progress on this path cannot be made without an increased awareness that all of us are part of one human family, united by bands of fraternity and solidarity."[1] You can find plenty of actual examples of this ethic on the planet: Scandinavia, say, or to a lesser degree any of the "welfare states" where people concern themselves with one another's . . . welfare. And it works. The World Happiness Report for 2018 found Finland the most cheerful country on earth, followed by Norway, Denmark, and Iceland. America came in eighteenth, "substantially below most comparably wealthy nations."[2] Even for Americans, though, it's not some odd, impossible concept. In 2017, amid Trumpist triumphalism, polls found that 61 percent of Republicans and 93 percent of Democrats wanted to "maintain or increase spending for economic assistance to needy people."[3] Ayn Rand has carried the CEOs, but not the rest of us. By and large, humans continue to believe in humanity.

And, luckily, we have two new technologies that could turn that belief into reality, two relatively new inventions that could, in our own era, prove decisive if fully employed. One is the solar panel, and the other is the nonviolent movement. Obviously, they are not the same sort of inventions: the solar panel (and its cousins the wind turbine and the lithium-ion battery) is hardware, while the ability to organize en masse for change is more akin to software. Indeed, even to call nonviolent campaigning a "technology" will strike some as odd. Each is still in its infancy; we deploy them, but fairly blindly, finding out by trial and error their best uses. Both come with inherent limits: neither is as decisive or as immediately powerful as, say, a nuclear weapon or a coal-fired power plant. But both are transformative nonetheless—and, crucially, the power they wield is human in scale. They don't threaten the game we've been playing all these years. Indeed, they threaten to make it more beautiful.

Before we discuss how we might best make use of these technologies, though, we need to engage the two most insidious ideas deployed in defense of the status quo. The first is that there is no need for mass resistance or government regulation because each of us should choose for ourselves the future we want. The second is that there is no possibility of resistance because the die is already cast.

Choice first. It is the mantra that unites people of many political persuasions. Conservatives say, "You're not the boss of me," when it comes to paying taxes; liberals say it when the topic is marijuana. The easiest, laziest way to dispense with a controversy is to say, "Do what you want; don't tell me what to do." So, if someone says, "I want to genetically engineer my child," many will defer to that person's choice. Jennifer Doudna, for instance, after a book-length discussion of her CRISPR invention, ends by writing, "I find myself returning again and again to the issue of choice. Above all else, we must respect people's freedom to choose their own genetic destiny and strive for healthier, happier lives. If people are given this freedom of choice, they will do with it what they personally think is right—whatever that may be."[4]

I've already explained the most obvious reason that this supposed freedom seems profoundly coercive: You can, with some effort, rebel against the ways you were raised. (Consider, say, the number of lapsed Catholics, despite all the combined and ferocious efforts of parents and nuns.) But you can't rebel against the genes implanted inside you: The choice your parents make in the fertility clinic will govern. In fact, if they get the dopamine just right, the idea of rebelling may never cross your mind.

There's something deeper at work here, too, though. This is the kind of choice that, *to a society*, isn't a choice at all. Once substantial numbers of people engage in genetic engineering, it will become effectively mandatory. Not by government diktat, but by the powerful forces of competition, as the possibility of improving your kids sets off a genetic arms race. The late MIT economist Lester Thurow posed the dilemma this way: "Suppose parents could add thirty points to their child's IQ. . . . And if you don't, your child will be the stupidest in the neighborhood."[5] This is one of those elections that happens only once: a fairly small number of people will make all the decisions for all time, in much the same way that a tiny clutch of people, by preventing us from addressing climate change, are making a decision that will stretch on deep into geological history. (This reasoning helps explain, by the way, why a large number of progressive feminists who support a woman's right to choose an abortion nonetheless oppose the right to tweak a baby's genes: in the second case, the effects extend to the entire society and pass down through the generations.) No one small group of people should get to make decisions like that by themselves. Such things should be decided (if anything should) by all of us.

The libertarian ideal of individual autonomy, which to one extent or another every modern human understands and cherishes, runs aground when the stakes get as high as ecological hell or posthuman meaninglessness. Much as I enjoy talking to Ray Kurzweil, he and his friends at Google shouldn't be allowed to unleash their vision on the world until we've all taken a vote.

If "Let anyone do what they want" is a flawed argument, then "No one can stop them anyway" is an infuriating one. Insisting that some horror is inevitable no matter what you do is the response of people who don't want to be bothered trying to stop it, and I've heard it too often to take it entirely seriously.

I remember, for instance, when investigative reporters proved that Exxon had known all about global warming and had covered up that knowledge. Plenty of people on the professionally jaded left told me, in one form or another, "Of course they did," or "All corporations lie," or "Nothing will ever happen to them anyway." This kind of knowing cynicism is no threat to the Exxons of the world—it's a gift. Happily, far more people reacted with usefully naïve outrage: before too long, people were comparing the oil giants with the tobacco companies, and some of the biggest cities in the country were suing them for damages. We don't know yet precisely how it will end, only that giving them a pass because of their power makes no sense.

Another example: seven years ago, some of us began fighting for fossil fuel divestment, and again we were told not to bother—if anyone did sell his stock, someone else would just buy it, and the world would roll on unchanged. But our little campaign became the largest of its kind in history—endowments and portfolios worth nearly $8 trillion have joined in—and it has clearly stung: recent academic studies have proved that it has helped move the climate issue to the fore and reduced the capital the fossil fuel companies can mobilize for new exploration. By 2018, after New York City and then Ireland had announced they were divesting, Shell Oil called the campaign a "material risk" to its business in its annual report, and Goldman Sachs analysts reported that the campaign had played a large role in devaluing coal shares. "It's inevitable" is a powerful argument right up to the moment when people decide not to let it sap their energy.

It's true that effective regulation of bioengineering or artificial intelligence will also be hard. CRISPR is so easy to use that high school biology labs can play around with genetic tweaking, and indeed, there are DIY gene-editing kits. (One entrepreneur has started including live frogs in his $159 mail-order kit, in the hope that people will stop experimenting on themselves.)[6] But hard is not the same as impossible: though scientists have edged up to the human germline, as of this writing, only Dr. He has actually crossed it, and the twin girls he modified in embryo actually led many scientists to call for more restrictions. His own clinical trials were shut down by the Chinese government. In most of the world where such work is possible, including all the European nations, heritable genetic modification is explicitly banned. Even if someone else decides to break the ban, a few designer babies are not like a few nuclear weapons: as the law professor Maxwell Mehlman once wrote, "If the number of enhanced individuals is sufficiently small, we may be able to ignore them."[7] A number large enough to matter would require investors to build clinics and chase payoffs, and that means the corporations involved would need to navigate the checkpoints of liability, of insurance, of large-scale financing. That is to say, they'd need to win the approval of the political system, which is not a gimme, at least not in a country like the United States, where strong majorities in every recent poll have serious reservations about such work: 83 percent of Americans told pollsters in 2015 that it was not appropriate to "change a baby's genetic characteristics to make the baby more intelligent."[8] Americans both right and left share this outlook: nine senators with 100 percent pro-choice voting records joined evangelical politicians to vote for a cloning ban.

It's not too late, not quite. Fossil fuel came to dominate our economy a century before we realized that global warming was a threat. That's one reason climate change has been so hard to bring under control. But human bioengineering and the most advanced forms of artificial intelligence haven't happened yet. Yes, Ray Kurzweil and Google have big plans and big power, but so far, Kurzweil's team is focused on auto-reply for Gmail. Yes, robots are scary, but "robots in a recent government-sponsored contest were stumped by an unlocked door that blocked their path at an outdoor obstacle course," the *Wall Street Journal* reported in 2017. ("One bipedal machine managed to wrap a claw around the door handle and open it but was flummoxed by a breeze that kept blowing the door shut before it could pass through."[9]) Some start-ups are currently employing humans to *pretend* they're robots: the software company Expensify had to hire humans to sort receipts because its robots couldn't handle the job, and the speech tech firm

SpinVox was hiring humans in foreign call centers to convert voice mails into text messages.[10] Our Alexa is significantly less competent than our mutt, who is unflummoxed by breezy doors.

None of this is to say that these technologies aren't coming, and soon—they are. But we've got a window, even if it's closing at exponential speed. It's still possible to imagine regulating AI, and we should. "Nobody likes being regulated," says Elon Musk, "but everything (cars, planes, food, drugs) that's a danger to the public is regulated. AI should be too." It's the "rare case," he told the National Governors Association, "where we need to be proactive."[11] With genetic engineering, it's crucial not to cross the germline and produce heritable alterations—that's as bright a line as 350 parts per million carbon dioxide in the atmosphere. With AI, the boundaries remain harder to draw: the search for a fail-safe switch to keep them from becoming too smart might be the most important engineering and policy task of our time. Some of that work is already under way: on Wall Street, where real money is at stake, people have proposed a variety of technological limitations to keep AI traders from crashing markets.

You'd need strong international regulations, too, across nations with very different notions about progress. The Chinese, for instance, seem significantly less worried about human genetic engineering—at least for somatic therapies, they're pushing ahead with human trials faster than anywhere on Earth,[12] though they reacted quickly to stop Dr. He after the news of his unauthorized designer babies. Meanwhile, China's AI start-ups now rival Silicon Valley's in size.[13] Vladimir Putin, touring a Moscow tech firm in late 2017, asked the CEO how long it would be before superintelligent robots "eat us." He added, "Whoever becomes the leader in this sphere will become the ruler of the world."[14] Elon Musk agrees—the competition for AI superiority, he posited, is "the most likely cause of World War III."[15] But that's exactly what diplomats are for (or used to be, when we had them). Since Hiroshima and Nagasaki, they've managed (knock on wood) to keep anyone else from exploding nuclear weapons in anger. They were aided in that task by the fact that all of us could envision the mushroom clouds that would come with failure, and that's not as easy with these other threats. Yet every nation on earth eventually signed on to the Paris accords to start dealing with climate change—and the U.S. was the first to drop out, which makes you wonder if China is really the main obstacle to international progress. And if we can't set up an international regime that lasts *forever*, that's not our responsibility. Our job is to keep the human game going through our time, and to pass it on.

Which means we should have a discussion, a long, deep, engaged discussion that involves all of us, everywhere. That's why I wrote this book, obviously, and perhaps it's the best argument for the set of digital tools we've evolved for global communication, flawed as those tools are. We should decide what we want. If it's to keep burning fossil fuels, or to craft designer babies, then, okay, we've made the decision. But we shouldn't pretend there's no decision to make, that it's simply inevitable. And we shouldn't leave the technicians to themselves, because if we do, they will simply press ahead, not because they're evil and only partly because they want to make a lot of money. Mostly they press ahead because it's what they do, and because there's great satisfaction in the work itself. The physicist J. Robert Oppenheimer, reflecting on building the atomic bomb, once said, "When you see something that is technically sweet, you go ahead and do it, and you argue about what to do about it only after you have had your technical success."[16] That technological focus is further distorted by the way science rewards innovators: as the pioneering AI analyst Eliezer Yudkowsky once explained, "Many ambitious people find it far less scary to think about destroying the world than to think about never amounting to much of anything at all. *All* the people I have met who think they are going to win eternal fame through their AI projects are like this."[17] We shouldn't let biologists and engineers decide whether and how to deploy these technologies, any more than we should let physicists decide where to drop nuclear weapons or let petroleum geologists decide how many wells to drill. They have special insight into *how* to do these things, but not into *whether* doing these things makes sense. When the effects of a decision will fall on the entire society, then entire societies should get to make the call.

But what if societies are just too timid or conservative? I talk often to technologists who say, "But what if we'd stopped innovating in 1800 or 1900 because we were worried about the effects?" "We romanticize humanity," Kurzweil insisted when we last spoke. "Read Thomas Hobbes, or even Dickens. Everyone lived in dire poverty; there was no social safety net. We have a moral imperative to continue on the pathway away from that because despite the substantial progress we've made, we still have suffering." Sure. But as every financial services firm has to tell you in its ads, "Past performance is no guarantee of future results." We're in a different world—in 1800 or 1900, the technologies on the horizon did not reasonably raise existential questions. The game was safe.

Innovation doesn't scare me (as you'll see when the discussion turns to solar panels), and it's not that we don't have problems left. (I've spent much of this book laying out the looming disaster of global warming and the deep harm done by our unprecedented levels of inequality.) The question is whether we can deal with those problems short of running the risks to human meaning that I've also described. I think we can. I think that if we back off the most crazed frontiers of technology, we can nonetheless figure out how to keep humans healthy, safe, productive—and human.

Again, though, not everyone agrees. Some harbor a deep pessimism about human nature, which I confess, as an American in the age of Trump, occasionally seems sound. Of all the arguments for unhindered technological growth, the single saddest (in the sense that it just gives up on human beings) comes from the Oxford don I mentioned earlier, Julian Savulescu. His proposal is important for our discussion because it unites the two halves of this book: in essence, he contends that the only way to solve global warming before it destroys our planet is to genetically alter human beings so that they become more altruistic and willing to make more sacrifices for the common good. He argues that we have "a moral obligation to overcome our moral limitations." People, he says, evolved to form groups of about 150 individuals, and to be violent to those outside their tribe. "We're far from perfect," he says, but "science offers us the opportunity . . . to directly overcome those limitations" by producing embryos with improved "intelligence, impulse control, self-control—some level of empathy or ability to understand other people's emotions, some willingness to make self-sacrificial decisions for other people," all qualities that "have some biological bases."[18]

Left to themselves, he insists, democracies can't solve climate change, "for in order to do so a majority of their voters must support the adoption of substantial restrictions on their excessively consumerist lifestyle, and there is no indication they would be willing to make such sacrifices."[19] Also, our ingrained suspicion of outsiders keeps us from working together globally. And so, faced with the need to move quickly, we should "morally bioenhance" our children, with the use of drugs or, more likely, genetic engineering, so they will cooperate. The changes will, Savulescu insists, be liberating: "Moral enhancement of a person does not restrict freedom; it rather extends it, by making the subject more capable of overcoming urges which counteract the doing of what is seen as morally good." It "hurts our pride," he theorizes, to "acknowledge our moral deficiencies," and yet we must do so, because the threats to our future must be answered by "morally responsible people."[20]

This whole scheme is roughly akin to "geoengineering the atmosphere" to prevent climate change—some people, having given up on taming the fossil fuel companies, want to instead pump the atmosphere full of sulfur to block incoming solar radiation. In both cases, it's an ugly workaround, based on the premise that we humans won't rise to the occasion. I hope that Savulescu seriously underestimates the power of both technology and democracy—of the solar panel and of nonviolence. As we shall see, we have the means at hand to solve our problems short of turning our children into saintly robots—which, in any event, wouldn't do a damned thing to solve climate change, given that by the time these morally improved youths had grown into positions of power, the damage would long since have been done. And I'm convinced he's wrong about people's selfishness presenting the main obstacle to solving climate change: around the world, polling shows that people are not just highly concerned about global warming, but also willing to pay a price to solve it. Americans, for instance, said in 2017 that they were willing to see their energy bills rise 15 percent and have the money spent on clean energy programs—that's about in line with the size of the carbon taxes that national groups have been campaigning for.[21]

The reason we don't have a solution to climate change has less to do with the greed of the great unengineered unwashed than with the greed of the almost unbelievably small percentage of people at the top of the energy heap. That is to say, the Koch brothers and the Exxon execs have never been willing to take a 15 percent slice off their profits, not when they could spend a much smaller share of their winnings corrupting the political debate with a broadside of lies and corrupting the political system with rolls of cash. If you wanted to "morally enhance" anyone, that's where you'd start—if there are Grinches in need of hearts, it's pretty obvious who should be at the head of the line.

But let's not win that way. Let's figure out how to solve the problems we face within the bounds of the game we've been playing all these millennia. Let's be, for a while, true optimists, and operate on the assumption that human beings are not grossly defective. Let's assume we're capable of acting together to do remarkable things.

20

You reach Moshono, on the outskirts of the Tanzanian city of Arusha, by following a badly rutted road that, at night, is both crowded with people and almost entirely dark. The town is about forty miles overland from the Olduvai Gorge, one of the places where anthropologists surmise humans first emerged, but until about a year ago, Lembris Andrea relied on a dim and flickering kerosene lantern to illuminate the home where he lives with his wife and two children. Thanks to a small solar panel on his roof, he now has five lights, including one by the corrugated tin front door, which he showed me several times. A "security light," he said. "To keep away criminals?" I asked. "Yes, crime is here, but also dangerous animals. Especially snakes, so it is good to have a light."

In the cocoa-farming village of Daban, in northern Ghana, I sat with several chiefs and elders on plastic chairs, discussing the new solar micro-grid that had gone into service the week before in their small settlement. It was blistering hot, as usual—we were six degrees from the equator—and one of the village leaders kept handing me the small plastic bags of water that are ubiquitous across West Africa. You just chew off a corner and then drink. I was grateful for the water, but it took me a good fifteen minutes to realize, in my clueless Western way, why the village elder was so proud. These bags were icy cold. This had not been possible until the switch was flipped on those solar panels the week before. For the first time ever in Daban, something *could* be cold.

Petite Boundiale, a rural village in the cocoa and rubber heartland of Côte d'Ivoire, is not prosperous: pretty much the only billboards on the nearby highway advertise machetes, and kids amuse themselves by pushing old motorcycle tires with sticks. But electricity had arrived, via solar panel, a couple of months before my visit, and so, a small crowd was gathered in the courtyard of Naore Abou's small house to watch the nineteen-inch flat-screen he'd set out on a bench. What does he like to watch? Football, of course. (The village is split between Manchester United and Liverpool, with a few Real Madrid supporters). But even more the National Geographic channel—that is, the electronic outlet of an institution that made its fortune showing pictures of remote African villages.

Solar power is a miracle, or at least close enough for our purposes. Like genetic engineering and artificial intelligence, it has roots in nineteenth-century science, but it came of age in the twentieth century and has reached true takeoff velocity in the twenty-first. Andrew Carnegie watched a Portuguese priest demonstrate an early "heliophore," which concentrated sunlight to produce temperatures of six thousand degrees, at the 1904 St. Louis World's Fair, a gathering that also saw the debut of the club sandwich, the hot dog, cotton candy, and the ice-cream cone. The sight stuck in Carnegie's mind—his steelworks having used by far the largest share of the world's coal supply, he predicted the best hope for humanity's future lay in "the sun-motor," whose "rays render the globe habitable, and may yet be made to produce power through solar engines."[1] Only in 1954, however, did the folks at Bell Labs manage to produce a working solar cell for generating electricity, and it was so expensive that at first its only use was in the earliest orbiting satellites. Taking advantage of the same learning curve that Kurzweil found with computing power, though, makers of solar saw the price begin to drop—indeed, the steady decline has taken solar power from a hundred dollars a watt in the 1960s to less than thirty cents a watt by 2018, making it the cheapest way to generate electricity across most of the world. No longer do you have to dig or drill for coal or gas or oil, and then ship it to a massive power plant, and then burn it at a high temperature, and then use the heat to spin a turbine, and then step the current up for transport through a far-flung grid, and then step it down again for home use. Now you can point a pane of glass at the sky, and out the back flows light and cold and information. That's Hogwarts-level magic.

I'd long understood the environmental benefits of the solar panel, of course. For thirty years environmentalists have been explaining that we need to replace fossil fuels with renewable energy in order to stanch the flood of carbon into the atmosphere. But I didn't really grasp the power of a solar panel to change lives until I got to rural Africa on a recent reporting trip. There are about as many humans living without electricity today as there were the day before Thomas Edison lit his first bulb, and most of them are in Africa. While Europe, North America, and South America are pretty close to fully electrified, and Asia is heading briskly in the same direction, the absolute number of Africans without power keeps increasing as population growth beats back the (minimal) efforts of the continent's utilities—a World Bank report that came out in May 2017 predicted that, based on current trends, there could still be half a billion Africans without power by 2040.[2] That's because conventional grids are expensive to build and maintain, and Africa is poor.

"The belief was you'd eventually build the U.S. grid here," said Xavier Helgesen, CEO of one of the most dynamic start-ups, Off-Grid Electric. "But the U.S. is the richest country on Earth and it wasn't fully electrified till the forties, and that was in an era of cheap copper for wires, cheap timber for poles, cheap coal, and cheap capital. None of that is so cheap anymore, at least over here." But solar—solar is suddenly cheap. And so, just as the spread of cheap cell phones a decade ago meant Africa could dispense with the need to wire landlines, so, too, solar may enable Africa to leapfrog at least some of the traditional ways of generating power. It's the ultimate in what the business gurus happily call disruption, and it's been a siren song for entrepreneurs with ambitions higher than the next Snapchat plug-in. Helgesen, for instance. Tall, with long lank hair, he could be the bassist in an indie band, but the Y Combinator T-shirt he's wearing gives the game away. He didn't actually do a stint at Silicon Valley's most famous incubator (his wife did), but that's his lineage, the same one that produced Airbnb and also the company that wants to embalm your brain so you can be digitally scanned and reimplanted in an android. The Y Combinator T-shirt reads, "Make Something People Want," which pretty much defines cheap solar power. Africans are desperate for electricity.

"This is how the solar revolution happens," Kim Schreiber, Off-Grid's communications director, whispers to me. "One hot sales meeting at a time." She and I, and the company's sales manager, Max-Marc Fossouo, are squeezed onto a bench in a courtyard outside of a hut in the Ivorian village of Grand-Zattry, which is not so grand. We're listening in as one of the company's newest salesmen, Seko Serge Lewis, tries his pitch. A couple of village dogs are growling and tussling nearby; a scooter rolls by with six people somehow on board. Next to us in the courtyard, a woman does the day's laundry in a bucket with a washboard—it's her husband the salesman is trying to woo. Right now he's showing him pictures on his cell phone of other customers in the village he's talked to already.

"That's to build up trust," says Fossouo in my ear. He's been providing a play-by-play throughout the hour-long sales call. "This customer is on a big fence. He's stuck in the trust place. And I'm pretty sure the decision maker is over there washing the clothes anyway."

Fossouo was born in Cameroon and went to school in Paris, but his real education seems to have come in the seven summers he spent in the United States selling books for Southwestern Publishing, a Nashville-based titan of door-to-door marketing. (Rick Perry is another alum; ditto Ken Starr.) "I did Los Angeles for years," he said. "'Hi, my name is Max. I'm a crazy college student from France, and I'm helping families with their kids' education. I've been talking to your neighbors A, B, and C, and I'd like to talk to you. Do you have a place where I can come in and sit down?'" All selling, he insists, is the same: "It starts with a person understanding they have a problem. Someone might live in the dark but not understand it's a problem. So, you have to show them. And then you have to create a sense of urgency to spend the money to solve the problem now."

After an hour of trying, Serge strikes out with that first customer—the man is flush now but worried that he won't be able to make the monthly payments down the road, in the lean stretch before the next cocoa harvest. "That's crap," Max whispers, pointing again to the wife bent over the laundry. "He loves this woman. He can move the world for her." So, as we move to the next small courtyard, Max takes over, so he can demonstrate his technique to the slightly abashed Serge. This prospect is a farmer and a schoolteacher, and we settle down in his classroom, which has a few low desks with slates—literal *shards* of slate—resting on top. Max quickly figures out that the man has two wives, and he starts sprinkling their names liberally through the conversation. "There's no pressure. It's okay. I don't want to sell you anything," he says, as they move through the steps familiar to anyone who's seen an infomercial. First Max has the man catalogue everything he's spending now on energy: kerosene, flashlight batteries, even the gas money for the scooter he borrows to travel to the next village when he needs to recharge his phone. Then he shows him what Off-Grid offers: a radio ("*plus fort*") and four lights, each with a dimmer switch. "Where would you put the lamp? In front of the door? Of course! And the big light in the middle of the room, so when you have a party, everyone could see. Now, tell me, if you went to the market to buy all of this, how much would it cost?"

The customer is resistant, but Max tries angle after angle. "You have to think big here. When I talked to your chief, he said, 'Don't think small.' If your kid could see the news on TV, he might say, 'I, too, could be president.'"

"This is great," the man says. "I know you're trying to help us. I just don't have the money. Life is hard, things are expensive, sometimes we're hungry."

Max nods, helpful. "What if I gave you a way to pay for it, so the dollar wouldn't even come from your pocket. If you get a system, people will pay you to charge their phones. Or, if you had a TV, you could charge people to come watch the football games."

"I couldn't charge a person for coming in to watch a game," the man says. "We're all one big family. If someone is wealthy enough to have a TV, everyone is welcome to it."

This hour also ends without a sale, but Max is not too worried. "It takes two or three approaches on average," he says. "You always have to leave the person in a good place, where he loves you stopping by. This guy wants to finish building his house right now; his house is heavy on him, but it won't be long." Indeed, even as we're talking, the first prospect wanders up to us again, asking for a leaflet and the phone number. His wife, he says, is very interested.

Part of me is happy that these two men have said no, at least for the moment. For an hour, for one of the first times in their lives, they've been in the position most Westerners occupy most of the time: that of a potential customer being wooed. It's a very different posture from being a supplicant to a government, and it comes with a certain undeniable power and dignity. And more often than not, the sale gets made, eventually.

Back in the regional capital of Soubré, I watch for a few moments as Off-Grid's champion Ivorian salesman winds up his third contract of the morning, on the way to nine for the day. He's tall and handsome in designer jeans and a dashiki-style shirt he had made in the company's colors. His name is Jean Anoh, but he's decided that that name is "too old, too soft." "Call me Stevens Ironman Never-Tired Killer," he says—

in between pitches, he's checking his phone for the constantly updated standings in Off-Grid's continent-wide sales championship and sending trash-talking WhatsApp messages to his Tanzanian competitors.

The overnight arrival of something as profound as electricity can't help but change communities, from how long people sleep to what they eat to the texture of village life. Some of the new electrical current gets used in ways everyone would approve of: I met five-year-old girls practicing their alphabet in exercise books by the bright new glare of an LED lamp. And when farmers have their phones charged up, they can get daily weather reports from Farmerline, a Ghanaian information service that uses GPS to customize the forecasts. "If a farmer puts fertilizer on the field and then it rains, he loses the fertilizer; it washes away," says Alloysius Attah, the young entrepreneur who founded the service. "The farmers say they can't tell the rain by themselves anymore. My auntie could read the clouds, the birds flying by, but the usual rainfall pattern has shifted."

But think about the percentage of your home's electric power that you use on things of utter virtue. It's no different in Africa. I met teenagers lying on the floor watching explosion-filled movies, and their younger brothers and sisters enjoying endless cartoons. I asked one young Tanzanian mother what her four-year-old watched, and the answer, through several layers of translation, was "N'klodin"—which is to say, Nickelodeon. "Our killer app is definitely the television," said Schreiber. "If the twenty-four-inch is out of stock, lots of people won't buy." One mother in a Ghanaian village explained that the TV "keeps the children at home at night, instead of roaming around."

"You must turn it off so they will study," a toothless old man told her, shaking his finger. But when I asked what was the most popular program, he and everyone immediately began laughing and nodding as one. "*Kumkum*," people shouted. *Kumkum Bhagya*, an Indian soap opera, airs every night from 7:30 to 8:30, at which point life grinds to a standstill. Loosely based on Jane Austen's *Sense and Sensibility*, it's set in a Punjabi marriage hall. "All the chiefs have advocated for everyone to watch, because it's about how relationships are built," said the chief of Kofihuikrom. "Too many marriages are on the downside."

At any event, this is what deep, rippling change looks like. The developed world has had two hundred years to absorb, more or less, what is now coming to rural Africa, belatedly but almost overnight.

The rapid spread of renewable energy across the developing world annoyed fossil fuel executives to no end. As they struggled to preserve their brands in the face of growing environmental clamor, the talking point they hit on was "energy poverty." The coal industry in particular cast itself as the solution to the lack of power in Africa. Peabody, for instance, the biggest of the coal producers, issued a "Plan to Eliminate Energy Poverty and Inequality," billing its product as "the only sustainable fuel with the scale to meet the primary energy needs of the world's rising populations." As one commentator explained, "Instead of a dirty symbol for pollution, the coal giant would refashion itself as a liberator of the world's poor, a way out of the darkness."[3] The talk about "energy poverty" was soon echoed by the Koch brothers, and by Exxon's Rex Tillerson, before he departed for his brief career in the Trump administration: pressed by environmental activists at an annual shareholder's meeting, he pushed back, launching into a brief sermon on energy inequality, concluding, "What good is it to save the planet if humanity suffers?"[4]

But, of course, humanity wasn't suffering from solar panels. Just the reverse—in communities that had been unlit, uncooled, and uninformed by fossil fuel for two hundred years, solar panels were turning on the energy overnight. By the middle of this decade, renewables were providing far more "new electricity access" than fossil fuels, and experts were calculating that meeting the UN's Sustainable Development Goals would require about 90 percent of new connections to come from the sun.[5]

You could perform the same solar trick in the developed world, of course, but here the complications are different. For one, people use far more power. It takes a roof full of panels to run my house, and even then, in the dark of winter, I rely on the larger grid—a good reminder that conservation and energy efficiency are as important as new supply. Meanwhile, the very existence of that grid lessens the impulse to put those panels up: almost all of us have access to highly reliable power all the time. And as we have seen, the utilities that provide that power don't want us to change—they've been willing to use their money and clout to dramatically slow down the trend toward renewable energy.

But we *could*. The sun shines everywhere, and when it doesn't, the wind is usually blowing. The latest studies, from labs such as Mark Jacobson's at Stanford, make clear that every major nation on earth could be supplying 80 percent of its power from renewables by 2030, at prices far cheaper than paying the damage for climate change. Jacobson's numbers are remarkably detailed: in Alabama, for example, residential rooftops offer a total of 59.7 square kilometers that are unshaded by trees and pointed in the right direction for solar panels. Want to know how much wind blows across the Mongolian steppe? They can tell you. As late as a decade ago, skeptics insisted that renewable energy would always be an outlier; it was just too expensive. But the engineers did what engineers do: they got better at everything, from predicting how hard the wind would blow to letting individual panels and turbines communicate directly with the grid. "After dropping 65 percent between 2009 and today," energy expert Dave Roberts wrote in 2017, "wind power costs could drop another 50 percent by 2020. That's pretty amazing." You don't need some magic technological breakthrough. Al you need is for "wind and solar to keep doing what they're doing—keep scaling up, keep improving, keep getting cheaper—at roughly the same rate they have been."[6] It's not that renewable energy is our only task. We also need to eat lower on the food chain, build public transit networks, densify cities, and start farming in ways that restore carbon to soils. But renewable energy may be the easiest of these tasks, especially since it's suddenly so cheap. The manufacturing process for solar panels has become so efficient that the panels pay back the energy used to make them in less than four years. Since they last three decades, that means a quarter-century of pollution-free operation.[7]

The most recent data come from Finnish and German researchers, and show the impact that rapidly falling prices for storage batteries have had on those calculations. These researchers found that by 2050, solar energy could provide 69 percent of our power and wind energy another 18 percent, with the rest coming mostly from hydroelectric dams. In the process, we'd create thirty-six million new jobs and the cost per megawatt hour would drop from the present eighty-two dollars to sixty-one dollars. The study's lead author, Christian Breyer, put it like this: "Energy transition is no longer a question of technical feasibility or economic viability, but of political will."[8] Other economists insist it would be cheaper and faster if there were some nuclear power in the mix, but the bottom line is fairly clear. If human beings wanted to, they could figure out how to extricate us from the climate mess by producing most of our energy from the wind and the sun. There's probably no single step that would do more to prolong the human game another generation, to pass the (solar) torch on to our kids and grandkids.

Yes, you'd have to build a hell of a lot of factories to turn out thousands of acres of solar panels, and wind turbines the length of football fields, and millions and millions of electric cars and buses. But here, again, experts have already begun to crunch the numbers. Tom Solomon, a retired engineer who oversaw the construction of one of the largest factories built in recent years, Intel's mammoth Rio Rancho semiconductor plant in New Mexico, took the Stanford data and calculated how much clean energy America would need to produce by 2050 to completely replace fossil fuels. The answer: 6,448 gigawatts. "In 2015, we installed sixteen gigawatts of clean power," Solomon says. "At that pace, it would take four hundred and five years, which is kind of too long."[9]

So, Solomon did the math to figure out how many factories it would take to produce 6,448 gigawatts of clean energy in the next thirty-five years. He started by looking at Tesla's big new solar panel factory in

Buffalo. "They're calling it the gigafactory," Solomon says, "because the panels it builds will produce one gigawatt worth of solar power every year." Using that plant as a rough yardstick, Solomon calculates that America needs 295 solar factories of a similar size to defeat climate change (roughly six per state), plus a similar effort for wind turbines.

We've mobilized at this scale once before, and it was the last time we faced what seemed like an existential enemy. After the attack on Pearl Harbor, the world's largest industrial plant under a single roof went up in six months, near Ypsilanti, Michigan; Charles Lindbergh called it the "Grand Canyon of the mechanized world." Within months, it was churning out a B-24 Liberator bomber every hour. Bombers! Huge, complicated planes, endlessly more intricate than solar panels or turbine blades—each one 1,225,000 parts, with 313,237 rivets. Nearby, in Warren, Michigan, the U.S. Army built a tank factory faster than it could build the power plant to run it—so it simply towed a steam locomotive into one end of the building to provide steam heat and electricity. That one factory produced more tanks than the Germans built in the entire course of the war.

It wasn't just weapons. In another corner of Michigan, a radiator company landed a contract for more than twenty million steel helmets, while the company that used to supply fabric for Ford's seat cushions went into parachute production. Nothing went to waste: When auto companies stopped making cars for the duration of the fighting, General Motors found it had thousands of 1939 model–year ashtrays piled up in inventory. So, it shipped them out to Seattle, where Boeing put them in long-range bombers headed for the Pacific. Pontiac made antiaircraft guns; Oldsmobile churned out cannons; Studebaker built engines for Flying Fortresses; Nash-Kelvinator produced propellers for British De Havillands; Hudson Motors fabricated wings for Helldivers and P-38 fighters; Buick manufactured tank destroyers; Fisher Body built thousands of M4 Sherman tanks; Cadillac turned out more than ten thousand light tanks. And that was just Detroit. The same sort of industrial mobilization took place all across America.

If (as the proposal for a Green New Deal envisions) we did something like that again, in the name of stopping climate change instead of fascism, we wouldn't have to kill a soul. In fact, we'd be saving great numbers of lives that would otherwise be lost not just from global warming but from breathing in the smoke of fossil fuel combustion. (The latest global data show that meeting the boldest climate targets set at the Paris talks would save 150 million lives, or roughly twice the number of people who died in World War II.)[10] And we wouldn't have to do it in mortal competition with the rest of the globe; this is a chance for cooperation on a new scale, as we pass technology back and forth. (At this point, China would clearly need to supply a good bit of the manufacturing muscle and expertise.) Instead of each of us being asked to sacrifice, we could each of us live in more comfortable and affordable homes, saving enough on fuel bills to finance the transition.

Does that sound too cheery, the idea of saving money as we move down this path? Rutland, Vermont, is a gritty place—it made the front page of the *New York Times* as a ground zero of New England's heroin epidemic—but I saw scenes there just as remarkable and just as moving as anything in Ghana or Tanzania. Sara and Mark Borkowski live in Rutland with their two young daughters, in a century-old, fifteen-hundred-square-foot house. Mark drives a school bus, and Sara works as a special-ed assistant. The cost of heating and cooling their home through the year consumes a large fraction of their combined income. Recently, however, persuaded by Green Mountain Power, the state's main electric utility, the Borkowskis decided to give their residence an energy makeover. In the course of several days, teams of contractors stuffed the house with new insulation, put in a heat pump for the hot water, and installed two air-source heat pumps to warm the home. They also switched all the lightbulbs to LEDs and put a small solar array on the slate roof of the garage. The Borkowskis paid for the improvements, but the utility financed the charges through their electric bill, which fell the very first month. Before the makeover, from October 2013 to January 2014, the Borkowskis used 3,411 kilowatt hours of electricity and 325 gallons of fuel oil. From October 2014 to January 2015, they used 2,856 kilowatt hours of electricity and no oil at all. They reduced the carbon footprint of their house by 88 percent in a matter of days, and at no net cost. If you multiply these kinds of small changes across many households, it pays off for everyone.

Green Mountain Power, for instance, was the first utility to subsidize its customers' purchase of Tesla Powerwall batteries. Two thousand Vermonters installed them, and when a savage heat wave hit in the summer of 2018, GMP was able to draw on the current they had stored away—which saved the state's rate payers half a million dollars in a single week, compared to the cost of buying outside power.[11] The same sort of change is possible everywhere, at almost any scale. In southern Australia in 2018, Tesla announced plans to build the world's largest "virtual power plant," covering the roofs of fifty thousand homes with solar panels that will be linked together to supply the grid. The panels are going up on public housing first, cutting residents' power bills by a third.[12]

Again, we're not talking about earth shelters in Aspen built of adobe and old tires by former software executives who converted to planetary consciousness at Burning Man. The Borkowskis' house couldn't be more ordinary: the girls' rooms feature *Frozen* bedspreads and One Direction posters, as well as two rabbits and a parakeet named Oliver. The family had no particular interest in the environment: "If it's not on the Disney Channel, I don't hear about it," Sara told me. The house sits in a less-than-picturesque neighborhood, in a town made famous for its junkies. Its significance lies in its ordinariness. If you can make a house like this affordably green, you should be able to do it anywhere, and if you can do that, then one of the great threats to the human game is partially defused.

21

Solar power is an interesting advance in that it's less powerful than the fuel it replaces, and in certain ways harder to use. Where coal and oil and gas can be gathered in a few places and shipped around the world, sun and wind must be collected from a million different locations and then shared across the grid; renewable energy is omnipresent but also diffuse, nothing like the concentrated package of chemical energy in a lump of coal or a liter of oil.

But these limitations also come with real and offsetting advantages. While some people will grow rich putting up windmills and solar panels, they won't make money on an Exxon scale because you can't charge for the sun. (That's why Exxon hates solar: you put up a solar panel and the energy comes *for free*, which to the corporate mind is the stupidest business plan ever.) The cash you spend for energy stays close to home; there's no way for the Koch brothers to become our richest and most powerful citizens simply by shipping fuel hither and yon.

In that sense, solar power could be a technology of *repair*, social as well as environmental. Even as it helps heal the atmosphere, it can help reduce that chunk of inequality that derives from the control of oil and gas deposits. In neither case can it do the job entirely: there are other sources of inequality, just as raising cows and cutting forests contribute to climate damage alongside power plants. But it's a start. And so, the fact that it doesn't represent a quantum leap forward in human power is a feature as well as a bug—it fits better with the human game.

You can say much the same for nonviolence, the other "technology" I want to posit as a practical hope. It works hand in hand with innovations such as solar panels, in fact, because if we are to build the political will to deploy renewable energy fast enough, we'll need a bulldozer for reshaping the zeitgeist. That's the job of movements.

Almost by definition, nonviolence is less immediately powerful than the forces (violence, coercion) it wishes to replace. (Unarmed protesters can always be shot, and many have been.) Violence is time-honored—it's the tool humans have used to resolve their differences over long millennia. When history needed to be unstuck, often the only crowbar for lurching forward was a revolution of some kind, fought with whatever weapons one could obtain. There's no need to view this history as abhorrent, any more than we need to think of burning coal in the twentieth century as some kind of crime. I grew up in Lexington, Massachusetts, and all summer long, I gave tours of the village green where the first battle of the American Revolution was fought. (Indeed, looking back, one sees that it was the first battle against the greatest empire the world had ever known.) The Minutemen were demanding real if imperfect democracy. One honors Captain Parker and his outgunned men.

But eight miles down the road from Lexington, in the town of Concord, and three generations later, another idea about resistance began fitfully to emerge from the complicated mind of Henry Thoreau. In 1846 he left the cabin at Walden Pond, where he was making his later-celebrated sojourn, and took a walk into town. He bumped into Sam Staples, Concord's constable and tax collector, who reminded him that he hadn't paid the annual poll tax required of all males between the ages of twenty and seventy. True, said Thoreau, he hadn't, because as an abolitionist, "I cannot for an instant recognize as my government that which is the slave's government also." So, he was led off to jail, and there he spent the night. His friend Ralph Waldo Emerson is said to have visited and asked why he was there, only to be asked in return, "Why are you not?" In any event, as Thoreau later wrote, he was thinking of solutions that went well beyond the simple democratic rule his New England ancestors had fought for:

> *Cast your whole vote, not a strip of paper merely, but your whole influence. A minority is powerless while it conforms to the majority; it is not even a minority then; but it is irresistible when it clogs by its whole weight. If the alternative is to keep all just men in prison, or give up war and slavery, the State will not hesitate which to choose. If a thousand men were not to pay their tax bills this year, that would not be a violent and bloody measure, as it would be to pay them, and enable the State to commit violence and shed innocent blood. This is, in fact, the definition of a peaceable revolution, if any such is possible.*[1]

Being an introvert and more than a little misanthropic, Thoreau was not about to organize anyone. His idea was quixotic enough that almost no one, including the fervent abolitionists around him, followed his lead. And the slavery question, of course, was settled only by a war as bloody as any in American history.

But if the world paid Thoreau little heed initially, time has amplified his idea. Tolstoy read him and wrote about him; through Tolstoy, Gandhi came to know of his essay, which he said in 1907 had been "written for all time. Its incisive logic is unanswerable." Gandhi's great "experiments with truth" occupied much of the next half century, and though they came too late to prevent the two world wars, they did eventually succeed in driving the British out of India. The work that the Minutemen had begun with muskets on Lexington's green, Gandhi and his followers finished with their courage at the saltworks in Gujarat—and then Dr. King adapted those techniques to take on some of the evils that the original American Revolution had institutionalized.

When I say "nonviolence," I do not mean only, or even mainly, the dramatic acts of civil disobedience that end in jail or a beating. I mean the full sweep of organizing aimed at building mass movements whose goal is to change the zeitgeist and, hence, the course of history. (Indeed, Gandhi made it clear that his satyagraha also included "constructive work" to build local economies. In his day, the key symbol was the spinning wheel, but now his old ashram at Sevagram boasts not only solar panels but a biodigester to make cooking gas from cow manure.) This movement-building differs from normal politics, from the daily fight for comparative advantage within the prevailing system and for the minor modifications of that system (tax cuts, say) that are in line with the existing, dominant currents of public opinion. Instead, I'm thinking of dramas such as the fight for suffrage, or against Jim Crow, or for gay marriage—each of which required a full-spectrum movement that stretched from the electoral to the illegal and was more focused on shifting culture than on winning narrow legislative victories. One of the finest theoreticians of nonviolence was Jonathan Schell, who, with his book *The Fate of the Earth*, had suggested that nuclear weapons, because they were so powerful, were rendering wars unfightable. In a subsequent book, *The Unconquerable World*, he advanced the idea further. Violence was increasingly dysfunctional, he wrote, and "forms of non-violent action can serve effectively in the place of violence at every level of political affairs." Or, more eloquently, it was the method by which "*the active many can overcome the ruthless few.*"[2]

I believe, as I've said before, that nonviolence is one of the signal inventions of our time—perhaps, if we are lucky, the innovation for which historians will most revere the twentieth century. Not everyone agrees. Indeed, to the most hard-nosed, it seems like so much faith-based mumbo-jumbo. The futurist Yuval Hariri said it was difficult to choose the twentieth century's greatest discovery. Antibiotics, perhaps? The computer? "Now ask yourself what was the influential discovery of . . . traditional religions in the twentieth century. That too is a very difficult problem, because there is so little to choose from."[3] His sneer is misplaced. True, nonviolence didn't emerge straight out of religion, and indeed, it sometimes subverted it—some of Gandhi's greatest campaigns were aimed at Hinduism's enduring caste discrimination. But mahatmas and ministers definitely led in developing this kind of resistance, and there is a spiritual insight at its core, one that traces at least back to the Sermon on the Mount. That's the idea of turning the other cheek, of taking on unearned suffering, of engaging our sympathy for the weak instead of our truckling admiration for the strong. Even in what seems like the very clinical world of environmentalism, mounds of research and data aren't ultimately decisive: the fight over climate change is ultimately not an argument about infrared absorption in the atmosphere, but about power and money and justice. Given that industry has most of that money and hence most of that power, it usually wins—unless, of course, a movement arises, one capable of changing hearts as well as minds.

Such a movement arose in the late 1960s, after the publication of Rachel Carson's *Silent Spring* and amid burning rivers and smog-choked cities. But it made little progress until Earth Day in 1970, when twenty million Americans (a tenth of the population) joined in demonstrations in every corner of the country. That unprecedented display of concern (and some subsequent electoral defeats for politicians tied to polluters) jolted Washington. With the usual balance of power upended, for a few years, major corporations lost one battle after another: Richard Nixon, no environmentalist, had little political choice but to sign the Clean Air Act, the Clean Water Act, the Endangered Species Act, and the other environmental laws still in effect today.

But success can sap movements. The organizations that had taken to the streets now retreated to big Washington offices, where they concentrated on lobbying. For a while, that strategy worked, because that first Earth Day had put enough juice in the battery to run a powerful motor for a decade or two. But that energy began to dwindle, and eventually the power of money reasserted itself. By the George W. Bush administration, the oil companies were back in control, able to scuttle any chance at progress on climate change. After Barack Obama's election, a Democratic Congress failed to pass even modest cap-and-trade legislation to fight global warming. And so, some of us decided that the time had come to try to rebuild the in-the-streets movement, so that we'd have a chance to win the fight as well as the argument.

Many people in many places have played many roles in that revival—especially, tellingly, those in the poorest places hardest hit by environmental change. For some of us, the vehicle was a small group called 350.org, which we formed in 2008 (the "we" initially being me and seven undergraduates at Middlebury College, in Vermont). We didn't concentrate on policy, but instead on mobilization, figuring that without a movement pressing for change, it was pointless to worry about precisely what change should look like. Our strategy was, frankly, ludicrous ("organize the world"), but beginner's luck has its place, and apparently there was an unfilled ecological niche. People around the planet were indeed worried about global warming, but they felt powerless against such a huge force; the mere act of gathering them together overcame some of that despair. Our first attempt to rally the globe, in 2009, saw 5,200 rallies in 181 countries, what CNN called the "most widespread day of political action in the planet's history." Most of those gatherings were small, but over time, the movement has grown to the point where we can put hundreds of thousands of people in the street. When we joined with others to fight the Keystone Pipeline, it helped set off a worldwide spate of battles that now tie down and complicate every fossil fuel infrastructure project. (The head of one of America's biggest energy lobbies has complained long and hard about the "Keystone-ization" of all the industry's plans.) Thousands in kayaks—"kayaktivists," of course—helped persuade Shell it didn't really want to drill in the Arctic; dozens of states and countries have now banned fracking; and some have gone so far as to stop new oil and gas exploration.

It's a movement now, and one increasingly led by kids, indigenous nations, communities of color. In the fall of 2018, a fifteen-year-old Swedish girl named Greta Thunberg staged a "school strike," sitting on the steps of Parliament instead of going to class on the theory that she couldn't be bothered if the government couldn't be bothered to care about the climate. Her action galvanized sentiment across northern Europe, and on the other side of the globe, Australian schoolchildren were soon on strike, too, and occupying the foyer of their Parliament. Meanwhile, in Britain an Extinction Rebellion movement had sprung up, staging civil disobedience actions to shut down traffic across London. In the United States, young people staged a sit-in at Congress to demand a special committee on a "Green New Deal" by early 2019 pollsters reported that 80 percent of Democrats and 60% of Republicans backed the idea, or at least the slogan." The Earth is running a fever, and the antibodies are starting to kick in. Which doesn't mean we've won. We haven't. The Koch brothers and the oil companies are holding on, thanks in part to Mr. Trump. We've scared them enough that they've begun to lash out—I've spent months with fossil fuel operatives following my every move with video cameras, thanks to the oil-funded efforts of the country's premier "opposition research" firm. After a remarkable display of indigenous unity at the Standing Rock protests against the Dakota Access Pipeline, Koch-funded legislators passed "anti-protester bills" in one statehouse after another, all designed to discourage that brand of dissent. (In Oklahoma, trespassing near "critical infrastructure facilities" now can get you ten years in prison.)[4] The same is true around the world, from Duterte's Philippines to Erdogan's Turkey to Maduro's Venezuela to Putin's Russia, where protest is often lethal. But the oligarchs face a fight at every turn. As Naomi Klein has said, if we can't get a serious carbon tax from a corrupted Congress, we can impose a de facto one with our bodies. And in so doing, we buy time for the renewables industry to expand—maybe even fast enough to catch up a little with the physics of global warming.

I recount all this not to boast—as I say, we're not winning, and in any event, I'm not much of a leader. (Having helped get things started, I've found it a relief and a pleasure to turn the spotlight toward the young, diverse, and remarkable organizers around the world.) Still, it has been a great privilege to see up close that, even against the biggest and richest forces on the planet, this technology of nonviolence can prove its power. The week in the fall of 2011 that civil disobedience against Keystone began at the White House, for instance, the *National Journal* reported on a poll of its 300 "energy insiders" on Capitol Hill and in the K Street lobbying shops. Ninety-one percent predicted that TransCanada, the company attempting to build the pipeline, would soon have its permit. But then 1,253 people went to jail, more than had committed civil disobedience about anything in decades. And then tens of thousands circled the White House—depending on your outlook, it was either a group hug or a temporary house arrest for Barack Obama. Before too many years had passed, polls showed a clear majority of Americans opposing this project. The pipeline's not built yet, and even if it someday goes in the ground, its chief legacy will be a widespread understanding that we shouldn't head down this path any longer. In the early summer of 2018, Pope Francis used precisely the language we'd pioneered in that fight: most oil, gas, and coal, he said, needed to "stay underground."[5] We'd begun to change the zeitgeist, which is the reason we'd gone to work in the first place.

It's possible to imagine a similar movement arising over, say, designer babies. In fact, such a campaign would in some ways be easier, because the proponents of such work aren't yet entrenched—there's not yet an Exxon equivalent, with a massive revenue stream and a harem of congressmen and senators. And activists would come to such a fight from both left and right, which means potential great power and also constant stress. I can't tell you what will trigger that movement—likely some development (a human clone? the next set of designer babies?) shocking enough to really capture people's attention—and I can't promise that it will win. If Google and companies like it mobilize on the other side, it will be a hard battle, probably won or lost on who is able to define what constitutes "progress." But it might well be the start of a larger movement in defense of the human.

Nonviolence is a powerful technology, despite the fact that we still know very little about it. Think about our understanding of military power: almost every nation on the planet has an academy or two devoted to the study of war, producing men and women who know everything about flanking maneuvers and close-air support. They join militaries better funded than any other part of our society. Our police are often heavily militarized as well. Visiting the encampment at Standing Rock was a reminder that state and local authorities, flush with surplus Pentagon gear, can seem almost indistinguishable from the armed forces. They had sound cannons and water cannons and vehicles that were, in essence, tanks; they wore the bulked-up tactical gear of the warrior. The oil companies hired security guards who came with snarling German shepherds.

And yet, all that firepower was almost powerless against the encampment that had gathered along the confluence of the Cannonball and the Missouri Rivers. In fact, the more force the oil barons deployed, the less it worked. The day they turned the dogs loose on peaceful demonstrators was the day that Standing Rock turned into a crisis for the White House, because people there knew what the pictures meant—they were a direct link to the iconic images of Birmingham and the civil rights movement. That Barack Obama was forced to enjoin the pipeline was a great victory; that Donald Trump bailed it out was a great, sad accident of history. But anyone who thinks that time is therefore on the side of the oil companies is reading history wrong. This movement will win (though, as we've seen, it may not win *in time*).

It will win in part because force has less purchase than it used to. Oh, it's still powerful, in the Dakotas, and in China, and in Russia, and in plenty of other places. But this new technology of nonviolence is challenging it. We're still early in the learning curve, and we lack a West Point or an Annapolis, but people around the world are trading lessons. The Serbian group Otpor! learns tactics from the overthrow of the nation's strongman Slobodan Milosevic, and it teaches those tactics to the young people mounting the Arab Spring; their successes and failures teach new lessons still. Over time, if we manage not to end the human game, these new ideas will continue to flourish, because they draw on precisely what is most human about us: creativity, wit, passion, spirit. None of these sounds like any match for money and weaponry, but ask the millions who rallied spontaneously to stop the president from separating families along the border. Sometimes this works.

Or ask the Nebraska rancher who managed, marvelously, to combine the crucial technologies of photovoltaics and civil disobedience. Bob Allpress raises cattle and alfalfa on a nine-hundred-acre spread that TransCanada Corporation wants to bisect with the Keystone XL Pipeline. He's been fighting the pipeline for years—and in 2017, he built a big solar array right in its path. If TransCanada wants to build the pipeline anyway, "not only would it have to invoke eminent domain against us, it would have to tear down solar panels that provide good clean power back to the grid and jobs for the people who build them," Allpress said.[6]

People are using the same tactic regularly now—in Nebraska, in Canada, in Australia, wherever a big new fossil fuel project is proposed. Some nuns recently built a chapel with a solar roof in the path of a pipeline. If you're an oil company, whom would you rather fight? A guy with a rifle is no problem; you've got access to all the rifles in the world. But a guy with some solar panels, access to social media, and a clever streak will drive you three kinds of nuts.

22

Imagine the last few hundred years of technological progress as a man spending an evening in a casino. He's had a remarkable run, one hot hand after another. There've been some losses along the way, sure, but he's always doubled down and made it back. Now, though, the bets are getting larger and larger, and his luck seems to be ebbing: if he doubles down again, he might lose it all. He sits and thinks a moment, and then, maybe, he takes his chips to the window and cashes them in, leaving with winnings that can secure the rest of his life.

Solar energy and nonviolence are technologies less of expansion than of repair, less of growth than of consolidation, less of disruption than of healing. They posit that we've grown powerful enough as a species, and that the job now is to make sure that that power is shared and controlled. They are, to use the first of several words that I wish we used more often, the technologies of *maturity*. They imagine a society interested more in economic and political maintenance and contentment than in exhilaration and extension.

In our current culture, we find the idea of maturing less exciting than the idea of growth because, I think, in our own lives, maturation is bittersweet. When we were young and growing, we could do and choose anything; no options had been foreclosed. Maturity—"growing up" as opposed to "growing"—means making choices: to commit to one person, one career, one community. In past times and places, that maturity was honored—look no further than the respect paid elders in more traditional societies, a respect reserved mostly for youth in our own consumer culture. But I'd guess that even a lot of Trump voters, in their heart of hearts, think most highly of those friends who have matured fully, which means those who have placed limits on their own behavior in the interests of the community. Such people find their fulfillment in working for others, in mentoring, in passing on—in behaving in precisely the altruistic ways that Ayn Rand and her followers so abhor. If we admire individuals for those traits, it's possible we can learn to admire societies for the same things.

Societies have already learned to accept some limits, and happily. For instance, I live on the edge of a federal wilderness area in the Green Mountain National Forest. For decades, that land has been set aside and protected: people are allowed in, but only as visitors. They wander the few trails, but most of the forest's tens of thousands of acres don't see a boot print from one year to the next; these are instead reserved for the turkey, the bear, the spruce. Yes, we gain from the arrangement, even financially: by cleaning the air and filtering the water, this wilderness provides "ecosystem services" that economists can measure, and those measurements assure us that the intact forest is a good bargain. But we forgo the quick cash, the growth, that would come from the liquidation of those forests into, say, wood pellets that could be burned in boilers to generate electricity. (This is, in fact, the current fate of too many American forests, especially in the Southeast.) And, of course, no particular person gets rich from this wilderness; it enriches us only as a society. So, setting aside wilderness is a statement that as a people we've reached the point where we can have a new ethic, in much the same way that getting married is a hopeful declaration that you've reached a place in your life where you want a new set of values to apply. And it's a popular statement—an "overwhelming majority" even of

Trump voters opposed the president's plan to roll back the size of the nation's protected areas.[1]

That doesn't mean that one needs to look back in horror at the days before such limits. There's much in our history to abhor, of course (slavery, sexism, and the way that Native Americans were treated, just for starters), but there's plenty to admire, too, or at least to look at with respect. Paul Bunyan, or the actual woodsmen on which he was modeled, managed to cut down most of the continent's forests with crosscut saws. That's harder work than I'll ever do, and it helped pave the way (literally) for the prosperity I enjoy. I bear no grudge to the Vermonters who came before me, whose stone walls you can find deep in the woods. They worked to build the world we know. But I also honor Vermonters such as George Perkins Marsh, the first American to postulate the ideas we now know as environmentalism. His careful nineteenth-century measurements of stream flows showed that cutting down forests was leading to spring flood and summer drought, unleashing great waves of silt. Given that there was no longer a New World for Americans to move on to, Marsh argued that we should set some limits on our behavior to keep this one healthy. And so, we did, beginning in the Adirondacks and Yellowstone, in an effort that spread around the world. Now 15 percent of the Earth's surface is protected. Societies are measured not just by the things they build, but also by the things they can bring themselves to leave alone: whales, bright-plumed birds, mountains, children kept safe from Dickensian labor.

People, *alone among creatures*, can decide to put such limits on themselves. None of these fights is easy; as I finish this manuscript, the Trump administration has just announced a new attack on the Endangered Species Act, on the grounds that it "impedes people's livelihood."[2] But in a world where algorithms are starting to take over, where Facebook and Amazon know us much too well, these self-imposed limits help keep us human. Our great literary conscience, the Kentucky farmer-writer Wendell Berry, said it best, long before anyone had heard of Cambridge Analytica:

Love the quick profit, the annual raise,
vacation with pay. Want more
of everything ready-made. Be afraid
to know your neighbors and to die.
And you will have a window in your head.
Not even your future will be a mystery
any more. Your mind will be punched in a card
and shut away in a little drawer.
When they want you to buy something
they will call you. When they want you
to die for profit they will let you know.
So, friends, every day do something
that won't compute. Love the Lord.
Love the world. Work for nothing.
Take all that you have and be poor.
[. . .]
As soon as the generals and the politicos
can predict the motions of your mind,
lose it. Leave it as a sign
to mark the false trail, the way
you didn't go. Be like the fox
who makes more tracks than necessary,
some in the wrong direction.
Practice resurrection.[3]

Wendell Berry is heir to a long countercultural heritage stretching back to the Buddha and running through the Christ, a tradition that incorporates people such as Thoreau and Gandhi, Dorothy Day and Ella Baker, and millions more women and men we've never heard of. This tradition celebrates limits, insisting that people are most fully human when they manage to restrain their own egos and desires. We've always paid it a good deal of lip service—America, for instance, is said to be a "Christian nation"—but now we might actually need that countercultural tradition to become more . . . cultural. The atmospheric concentration of carbon dioxide is a practical argument for what has previously been a moral stance; the periodic table is pointing us in the same direction as the Hebrew prophets.

And it's not just sages and gurus and cranks who have imagined such a thing. Perhaps, actually, it's best not to dwell on them, Jesus being literally a tough act to follow and Thoreau not the kind of guy you can imagine with a family to care for. Let's even tone down the language: *maturity* is perhaps a little stern and parental. Instead, let's add another word to our lexicon: *balance.* After forty years of libertarian dominance in our politics, ever since Ronald Reagan won by insisting that government was the problem and Thatcher by declaring that there was in fact no such thing as society, it's hard for us to see quite how lopsided our politics has become. The percentage of Americans who remember the New Deal grows tinier each day, and even Lyndon Johnson's Great Society seems from a different age. But we need to remember, because it's with such laws that human solidarity becomes everyday instead of exceptional.

You can't spend your entire life building movements—almost by definition, they burn bright and then burn out. (This is why greed usually wins, of course: the Kochs are at it 24/7. If a high-priced lobbyist succumbs to cirrhosis after the millionth cocktail reception, they just hire a new one.) So, we need structures that make fraternity real and relatively easy: labor unions, voting rights, a social safety net. Maybe state banks, like the one in North Dakota. Truly public utilities. These are not bizarre, Communist ideas. You can find examples of them around the continent and around the world, from the municipally owned internet service in Chattanooga (top-ranked by *Consumer Reports*) to the publicly owned nonprofit football team in Green Bay, Wisconsin (top-ranked by Cheeseheads). In Germany, 850 local cooperatives control much of the renewable energy that increasingly powers the country, most of them financing their operations from one of the country's thousand cooperative banks.[4] Saying "We need balance" is not the same as saying

"The economy's not important and we should live on craft beer and good vibrations." A quick transition to renewable energy would employ millions around the world by every estimate—millions of people, not robots, as clambering onto your roof and installing solar panels remains a high-skill, high-judgment job. When far more young people tell pollsters that they identify with socialism more than with capitalism,[5] they don't mean they want to live in North Korea; they mean they want a fair chance, not the loaded system they've inherited. Again, solidarity doesn't require saintliness. It requires the institutions that the antigovernment right has been trying to dismantle for decades.

Scale is the third and final word that seems crucial to me. If the only things you wanted in the world were efficiency and growth, then you'd scale things up—and we have: large corporations, large nations. But we've reached the point where size hinders as much as it helps, where it reduces the many ways the human game might be played down to just a few. Some of this has happened naturally. As humans explored the globe and ran into one another, the number of separate games dwindled. Mexicans no longer have a world to themselves: now the rest of us have corn and chili peppers, too—for which, many thanks. But when NAFTA turned the glut of corn into destitution for Mexican farmers, it showed how scale might become too big.

Protectionism is a vulgar word for economists because it's inefficient, but *inefficiency* is often just another way of saying that you serve more than one end. Amazon is incredibly efficient—I can have something that I may or may not need at my doorstep tomorrow—but when it puts actual stores out of business, it sacrifices the other services those actual stores provided: "gossip, help for old people, surveillance of the street."[6] Cargill and Archer Daniels Midland are incredibly efficient—I can buy food for very little money—but when they put local farmers out of business, we lose rural communities, pastoral landscapes, agricultural diversity. (Also, we grow plump on corn syrup.) We find out what those benefits are worth only when they evaporate, and even then, the losses register mostly unconsciously—you don't know what you've got till it's gone, by which point you're usually pretty well accustomed to the replacement.

This is why I think it's useful that both nonviolence and solar panels nudge us, at least a little, toward a smaller-scale world less obsessed with efficiency. Movements are often reminders about precisely the things that have been sacrificed in the name of efficiency. We can thank organizers for eight-hour days and five-day weeks; for Social Security and minimum wages; for the fact that we have cleaner air in our big cities and cleaner water in our rivers. Really, we can thank them for all the things the Koch brothers don't like. Solar power accelerates this transition. Because it comes from everywhere, it gives everywhere the chance to provide for more of its own needs. Instead of bleeding capital off to Saudi Arabia or Texas, we can increasingly supply one of our most important commodities close to home. We'll have to fight to make sure this happens—that communities control their local energy sources, and that those sources are developed with everyone's interests in mind—but at least it's a possibility. Home, community, is the ground on which we can actually play the human game, and it is false efficiency to undermine it.

I think I'm not alone when I say I find America increasingly perplexing: three hundred million may be past the population size at which any of us can feel fully at home or completely responsible. I try to imagine Donald Trump coming to a town meeting in my small community, the first-Tuesday-in-March gathering where we vote on the budget for the year and discuss community business. His foul mouth and obvious disdain for detail would mean that no one would pay him much mind; if he kept up his ranting, he'd be asked to sit down so that the rest of us could do the necessary work of making sure there was money to buy sand for the road crew and of figuring out if the roof on the town office had another year of life in it. Donald Trump, I think, would have had a hard time being elected a mayor or a governor, because the damage he'd have done would have hit too close to home. But given the size of America, people could vote for him for president on the theory that he'd "shake things up," reasonably confident that they wouldn't be hit by the falling pieces.

What I'm trying to say is what worked in the past doesn't automatically work in the future. At one point, growth provided more benefit than cost. Light regulation spurred expansion. Larger scale offered efficiencies that made us richer. Fine. You want your child to grow—if she doesn't, you take her to the doctor. But if she's twenty-two and still shooting up by six inches a year, you take her to the doctor, too. There's a time and a place for growth, and a time and a place for maturity, for balance, for scale. And the risks we're currently running, the risks I've spent this book describing, suggest that that time is now. In fact, the damage we can already see, from soaring temperatures to soaring inequality, should tell us that our goals need to fundamentally shift: toward repair, toward security, toward protection.

The overarching goal is to keep the human game going. To return to our casino metaphor, what if we collected our winnings from the last few hundred years and then decided we'd take a rest, play some lower-stakes hands for a while. Perhaps our job, at this particular point in time, is to slow things down, just as basketball teams do when they're ahead. If we don't screw up the game of being human, it could last for a very long time; compared to other species, we're still early in our career. (Consider horseshoe crabs, 445 million years old, so old that their blood is copper-based. Now *that's* a good long run.) And calculate the risks: if we manage to screw up the human game, through some combination of environmental destruction and technological usurpation, we prevent the hundreds of billions of perfectly interesting and amiable lives we could otherwise expect in the eons ahead. We also waste, in some sense, the work that every poet, philosopher, and scientist has done over the last ten millennia. Given that there's no finishing line to the human game, no obvious goal toward which we are racing, then why exactly are we so intent on constantly speeding up?

Indeed, we're surrounded by signals flashing amber, telling us to slow down. That's how to read the graph of a rising temperature, or the astonishing data on skyrocketing inequality. There are subtler indications, too, that we're at or near the top of a curve. To take the least important first, the performance of our athletes has begun to plateau, as records get harder to break by even tiny margins: the 2000s were the first decade in a century of measuring when no man ran a faster mile; the 2010s so far are the second. As the sportswriter Clint Carter points out, at least a dozen track-and-field events, including the 3,000- and 1,500-meter runs, haven't seen a single new record in more than two decades. "The long-jump record has gone untouched for 27 years; the shot-put record for 28 years. Both discus and hammer-throw records were established more than 30 years ago," he observed in the summer of 2018.[7] Indeed, in some sports, times have begun to slow as authorities have managed, at least for the moment, to crack down on dopers. (It takes authentic humans longer to climb the Alpe d'Huez than it took Lance Armstrong.)

It's not just elite athletes who have hit a plateau; at least in the Western world, almost all of us seem to be stalling out. Recent studies seem to show that while "the twentieth century was an unprecedented period of improvement for human capabilities and performances, with a significant increase in lifespan, adult height, and maximal physiological performance," the data now show "a major slowdown occurring in the most recent years." Just as we're no longer dramatically increasing our crop yields the way we did after World War II, we've also stopped growing much taller; the rate at which our lifespans lengthen has also begun to slow.[8]

This slowdown seems to be affecting our brains as much as our bodies, which will come as a shock to many. Steven Pinker devoted a sizeable chunk of his optimistic book *Enlightenment Now* to demonstrating that IQs were surging. "Could the world be getting not just more literate and knowledgeable, but actually smarter?" he asked with trademark perkiness. "Amazingly, the answer is yes. IQ scores have been rising for more than a century in every part of the world, at a rate of about 3

IQ points per decade." This Flynn effect (named for its discoverer) provided what Pinker called a "tailwind in life," "a gateway to compassion and ethics."[9] So that makes it tough to read the new data that emerged in 2018 showing the Flynn effect now running in reverse, with IQ "hitting its peak for people born in the 1970s and significantly declining ever since." A review of seven hundred thousand IQ records in Norway showed that IQs were now dropping by seven points per generation, and the same kind of declines were seen in the six other nations studied. "It's not that dumb people are having more kids than smart people," said one of the researchers. "It's something to do with the environment, because we're seeing the same differences within families."[10]

Taken all together, the results suggest that instead of dreaming about utopia, we should be fixated on keeping dystopia at bay. And a team of scientists who performed the biggest global meta-study of this data on human performance ended their report with precisely this suggestion: our task now should be to somehow *maintain the gains of the past.* "Care should be taken," these researchers concluded, "to prevent regression even if remaining close to upper limits may become more costly. This aim will be one of the most intense challenges of this century, especially with the new pressure of anthropocentric activities responsible for deleterious effects" on our health and well-being.[11] That is to say, we're at something like peak human right now, and it would be a worthy task to try to stay there—to spread the benefits of the last hundred years in diet and public health across geography and class, and to try to ward off the side effects of twentieth-century progress before it compromises twenty-first-century lives.

Clearly there are plenty of places that need to catch up, whole continents full of people who haven't benefited much from the long, hot streak in the casino. Such places supplied much of the labor and raw material that made those winnings possible, often not by choice; and they're paying the early price for our overreach as their oceans swell, crops wither, and forests burn—or as their jobs disappear with the rise of robots. And even as we've gotten taller, healthier, and longer-lived, there are of course people who still get sick. And all of us still die.

These inequities can be used to promote the current way of doing business: think of all those coal executives, for instance, who developed a sudden interest in "energy poverty" when the developed world began to balk at burning more of their product. Even those philosophers who think we're solving most of the world's troubles believe that we should double down on business as usual to mop up the remaining woes: Pinker, for instance, refuses to brook even momentary slowdowns or limits; his prescription "for today's bioethicists can be summarized in a single sentence: Get out of the way."[12]

It's true that one way to deal with our remaining troubles, with the inequality and the remaining underdevelopment, is to try to amp up the growth machine again: If we cut the taxes of the rich, it will theoretically generate prosperity. If we unleash industry from environmental rules, it could produce jobs. If we pay no attention to the atmosphere, coal-fired power could conceivably bring prosperity to Africa, not to mention West Virginia. These are the promises of Trumpism and Kochism, and they're not so far removed from the idea that if we plunge full speed ahead with artificial intelligence, it, too, will "grow our economy." This approach has the advantage of being just like the past: it's obviously easiest to keep doing what we've been doing. As the French journalist Hervé Kempf observed, growth "creates a surplus of apparent wealth that allows the system to be lubricated without modifying its structure."[13] But as this book has pointed out, that growth now comes with enormous levels of risk. Indeed, it risks ending the game of being human. Even with all that lubrication, the gears have begun to grind.

And so, it makes sense to remember that those who helped define our current worldview have themselves imagined other possibilities. Adam Smith, who with *The Wealth of Nations* fired the gun that set off the race we're still running, nonetheless predicted that the time would come when "a country which had acquired that full complement of riches which the nature of its soil and climate, and its situation with regard to other countries, allowed it to acquire; which could, therefore, advance no farther, and which was not going backward."[14] This stationary state was the inevitable destiny of societies, he believed, even if no one had gotten there yet. The great philosopher and political theorist John Stuart Mill, revered by many libertarians for his classic essay *On Liberty*, could scarcely wait for such a steady-state economy. "A stationary condition of capital and population implies no stationary state of human improvement," he wrote. "There would be as much scope as ever for all kinds of mental culture, and moral and social progress; as much room for improving the art of living, and much more likelihood of it being improved, when minds ceased to be engrossed by the art of getting on."[15] And, in living memory, it was John Maynard Keynes who hoped that "the day is not far off when the economic problem will take the back seat where it belongs, and the arena of the heart and the head will be occupied or reoccupied, by our real problems—the problems of life and human relations."[16]

Which is why I keep flashing back to one of the most interesting people I met in my travels for this book. Her name is Nicole Poindexter and she's African American. Raised in Texas, where her father was a surgeon, she was schooled at all the right places: Yale, Harvard Business School. She spent her time on the trading desks of the investment banks, playing with the derivatives that helped crater the economy, and then at Opower, a software platform for utility customers that was acquired not long ago by the tech giant Oracle. "I was an early employee there, and eventually it went public. I love what they're doing, but there was a nagging sense that that was not what gave me purpose," she said.

Poindexter and I were sitting in the back of a car bouncing along a dirt road near the Northern Ghanaian city of Kumasi—Ashanti country, hot as heck. A long way from Harvard. "I saw this one video; it was during the Ebola crisis," she told me. "People were living in pre-industrial conditions—I mean, they were pushing with their feet to power a forge. There was a lot of coughing in the background, and I was thinking, 'That's someone with Ebola.' But it wasn't—it was from the smoke in the room from the fire. So, that is unacceptable to me, not when we in our world have this abundance. And I put it together with the energy stuff that was in my mind from Opower, and I got on a plane."

Poindexter had a framework in mind different from that of most of the entrepreneurs I met. Where they were focused on selling individual customers single-panel systems, her idea was village-scale solar microgrids. She wanted to build small solar arrays on the edge of rural towns and then wire the huts, as if she were a miniature Con Edison. The model requires more capital up front for the company, and hence more risk, but it also means you can provide more power, enough to think beyond lights and televisions and toward moneymaking businesses. The financial model imagines that customers will start by using almost nothing, just a hundred kilowatt hours a month, but it assumes that in the course of a decade, as they figure out what to do with that electricity, they will end up using a thousand kilowatt hours.

"What's your evidence for the assumption that people will increase their usage like that?" I asked.

"My evidence is just, oh, everything in history," she said. "If people have access to power, they will make use of it."

The numbers worked—on paper, anyway. "One day I built the model," Poindexter explained. "I costed everything out, ran all the numbers. And at the end of the day, it showed that a system for a small village could provide two thousand dollars in profit. And I thought, that's the most amount of work I've ever done to get two thousand dollars. But it *was* making money." In fact, if people do increase their use as expected, she says her investors will get "a fifty-percent return, unlevered."

With a colleague, Joe Philip, who is Indian American and had been working at the renewable energy start-up SunEdison, Poindexter put together a small round of financing in 2015, and they started their first project in the Kumasi region, under the Black Star Energy label. (Check out the Ghanaian flag and you'll get the name.) None of it was easy. American-style smart meters, at fifty bucks a pop, were way too expensive, for instance, so Philip and his team built their own, at a buck apiece, with chips ordered from Amazon. Kumasi, the regional capital, where Black Star's headquarters was located, had grid power as unreliable as everyplace else in Ghana, making the office almost impossible to work in. "You'd get twenty-four hours on, then twelve off," said Philip. "Every time you came back to your apartment it would be off." But that, of course, only increased his and Poindexter's resolve, by reminding them what life was like for their potential customers. "If you don't have lights, you're always rushing," Poindexter said. "You're rushing home from the field to make dinner before it gets dark. Everything has to happen in a twelve-hour day." And so, they worked quickly to wire their first community.

Our car was now bumping to a stop outside one of these first-to-be-wired villages, Kofihuikrom. We got out and inspected the small, fenced-in array of solar panels and then walked to the most prominent building in the settlement, a cement-block clinic with a big poster on one wall showing Nelson Mandela talking about tuberculosis. The clinic's director was there to shake our hands. "I always had to store vaccines in different villages—in a different district," he said. "No refrigerator. Now—now I can make ice packs for people. When I came here, we were using flashlights to see patients. That had to stop. We were trying to deliver babies with flashlights. Not the good kind you wear on your head, but holding it in your mouth to see. Now we have night hours." It used to be hard to hire a nurse: "People didn't want the posting here. But the new nurse came to spy before he took the job. When he saw we had power, he said, 'Okay.'"

A few feet away, cacao nuts were drying on screens—we had arrived just at the April harvest. This is a *very* poor place. Poindexter guessed that the average income per household was about three dollars a day, and the average household had five members. So, the village needed power not just for light, but for moving farther up the supply chain. "I really like chocolate," she said. "And I just bought some at the airport in Amsterdam for eighteen dollars a pound. And when I do the numbers on the back of an envelope, I think the farmers here, my customers, make about one cent a pound. But if they're doing the winnowing, the roasting themselves—maybe a dollar a pound someday? Plus, then you're shipping cacao nibs, not all the water that's in this nut."

To wander through these newly electrified villages is to understand an awful lot about why the twentieth century was so amazing: its delayed arrival here lets one sense what it was like in America in the 1930s, as rural electrification rolled across the countryside; or in China in the 1990s. But to imagine power arriving without pollution is a nice twist: These communities in rural Ghana aren't getting the oldest, cheapest technology. They're getting the newest, cheapest technology, so new and so cheap that families who currently make three dollars a day can afford it. And this isn't aid; it's a business. When we meet Poindexter's customers, they're grateful but not obsequious, and she's gracious but not sentimental. They ask about credit for refrigerators, about the possibility of streetlights, about other appliances. There's not a Luddite in sight, nor a romantic. "I'm not a socialist," said Poindexter. "I don't think humans are wired that way. But I also think extractive capitalism has run its course."

How *are* humans wired? Where's the sweet spot, the balance, the right scale? "On my very first trip to Ghana, I was certain of many things," said Poindexter. "Like, these were communal societies; we can have one meter for each village. 'No, no,' everyone said. 'That won't work. We'll fight over who used the most power.' So, we have individual meters." Score one for Ayn Rand. But Poindexter's system is nonetheless based on communities. "If we have a village with a hundred households, we need sixty of them to sign up before we go forward. Every individual needs to choose whether they want it, but they also have to work together as a community. We meet with the chiefs first, we sign a memorandum of understanding with the chief or the queen mother. We're a utility. The unit is the community."

It's also the community that changes, doubtless some for the better and some for the worse. In Côte d'Ivoire, I met a farmer who said his life was much improved by the new power. "In the old time, you had to go outside and talk. Now my neighbor has his TV, I have my TV, and we stay inside"—which to me sounds like the first sad step toward *Atlas Shrugged*, but people have a right to figure this out on their own. I sat with a chief in the Ghanaian village of Daban, sipping ice-cold water and listening to him talk about the advent of power. "On the third day we had it, all the youths went to the city to bring back a sound system," he said. "We played music all night. Even that old man there, he has been playing music."

23

Which brings me back to a point I've been making in the margins all along. The human game is a team sport.

Or, at least, so it seems to me. If the antigovernment conservatives are right instead, and individuals are all that really matter, if "there is no such thing as society," then we do not stand a chance. We won't be able to mount a real common effort against climate change. We'll stand and watch, slack-jawed, as the latest inventions roll out of Silicon Valley.

But I'm pretty sure they're wrong. The human project has long been a group effort. We're born big-brained but unformed and vulnerable. It took a tribe, a band, a clan, a community to raise humans to adulthood. We hunted together in groups. With our complex language, we're able to gossip, to keep track of one another. And everything we learn about the human animal, now that we can stick people in MRIs or analyze their hormones, leads us to think that we're still not that far removed from the creatures who sat on the floor of the savanna picking lice from one another's fur.

In 2018, the Centers for Disease Control released startling new statistics on suicide in America: since 1999, it has gone up by 25 percent "across most ethnic and age groups."[1] Those are astonishing numbers, and hard at first to explain: during these same years, far more people have been able to find treatment for depression and anxiety. Clay Routledge is a behavioral scientist at the state university in North Dakota, which saw a 58 percent rise in suicides, the largest in any state. Humans, he wrote recently, require not just food and shelter but "meaning and purpose." We can't easily manufacture it on our own—"the psychological literature suggests that close relationships with other people are our greatest existential resource"—but we live in a world where families form later in life, if at all; where the religious institutions that once brought us together have begun to wither; and where screen-bound people "are less likely to know and interact with their neighbors." This is hideously bad news because "studies have shown that the more people feel a strong sense of belongingness, the more they perceive life as meaningful"[2]—meaningful enough for them to keep on living it.

Some of these modern woes might yield to political changes. Because Scandinavian countries provide the child care that makes it easy to be a parent, for instance, their citizens form larger families. Quite likely, progressives should pour less scorn on churches, if only because they provide a place for people to gather. The basic question of whether society really matters: that could transcend left and right.

It doesn't, however, transcend politics. The antigovernment impulse currently runs our world. It's expressed by all those cabinet officials who keep *The Fountainhead* on the bedside table, all those billionaires who gather with the Koch brothers to figure out the course of our politics, all those Silicon Valley moguls who want nothing standing in the way of their next inventions. These are people who, at some level, hate the idea of society, who organize campaigns against public transit, who try to dismantle public schools and national parks, who instinctively head for the gated enclave. I don't think their rule will last forever, but as I've said, they currently possess a savage leverage, perhaps power enough to end the human game. Certainly, they're trying their best. The endless efforts to gerrymander districts, suppress voting, race-bait, gin up cynicism in our politics, confuse us about issues such as climate change—these are nothing more than efforts to weaken society so it can't exert power over its most dominant individuals. Polling shows that "the poor are now democracy's strongest fans, the rich its biggest skeptics."[3]

Another way of saying this: one reason that some powerful people like robots is precisely because they come without the human impulse toward solidarity—they didn't need a society to rear them; they are immaculately self-possessed. Andy Puzder, Trump's first candidate for secretary of labor, devotes "much of his free time" to reading Ayn Rand. His day job was running the Hardee's and Carl's Jr. fast-food emporiums, and in that role, he bitterly opposed raising the minimum wage; people who wanted fifteen dollars an hour, he said, "should really think about what they're doing." Instead, he yearned for a future of greater automation at his chains, because robots are "always polite, they always upsell, they never take a vacation, they never show up late, there's never a slip-and-fall or an age, sex, or race discrimination case."[4]

Something else about robots: a nonviolent campaign would have no effect on them. They'd view the Montgomery Bus Boycott as an illogical exercise. An AI could beat a Gary Kasparov, but it would blink uncomprehendingly at a Colin Kaepernick. The appeal to human solidarity, to fellow feeling, reaches its limits at the borders of consciousness. So, we'd best get started soon if we're going to get started at all.

Epilogue

Grounded

I was eight years old in July 1969, and so, almost from the start, my dawning sense of the outside world included the larger universe. I watched Apollo 11 from hours before liftoff to the end of the mission a week later, switching off the television only for occasional bouts of parentally mandated sleep and for trips to the backyard, where I would stare at the moon in wonder. I memorized all the NASA acronyms (LEM, for "lunar excursion module"; EVA for "extravehicular activity," which is to say, walking on another heavenly body). I recited the countdown out loud, over and over: "T-minus twelve, eleven, ten, nine, ignition sequence started, six, five, four, three, two, one, zero. We have liftoff. *Apollo Eleven* has cleared the tower."

So, in the spring of 2018, forty-nine years later, it was a journey back into my deep innocence to stand on the roof of the VAB (Vehicle Assembly Building, of course, the tallest building outside an urban area and the structure with the largest doors on planet Earth and the biggest painting of an American flag) and stare out across the Cape Canaveral scrub in the hour before dawn as a rocket owned by Elon Musk prepared to hurl itself toward the International Space Station. When the moment came, it was as I'd always imagined: the clouds of steam as gas vented, then the immensely bright column of flame erupting. For a second, nothing seemed to happen—until, with remarkable slowness, the rocket began to rise, the grip of gravity yielding to the force of its engines. As the Falcon 9 began to accelerate, a great ripping sound caught up to the sight; ever faster, the ship rose into the paling sky, Roman-candling through the clouds for six minutes before it finally disappeared from straining eyes.

It is the most awesome technological spectacle humans have produced, universally seductive. Even to Ayn Rand. She covered the launch of the moonshot for her magazine, *The Objectivist*, and though she of course insisted the government had no business funding such things, she allowed herself to be overcome anyway: "What we had seen, in naked essentials—but in reality, not in a work of art—was the concretized abstraction of man's greatness . . . that a long, sustained, disciplined effort had gone to achieve this series of moments, and that man was succeeding, succeeding, succeeding." Better yet, she wrote, Neil Armstrong didn't ruin his great moment on the moon by talking about God, "did not undercut the rationality of his achievement by paying tribute to the forces of its opposite; he spoke of man. 'That's one small step for a man, one giant leap for mankind.' So it was."

Rand would like the current space program even more. President Trump has proposed zeroing out the budget for the International Space Station, meaning that much of America's reach into space will need to be funded by the band of tech billionaires who have seized the opportunity. On this day, it was Musk's company SpaceX, but the flare of rocket engines also illuminated the vast hangar of Jeff Bezos's Blue Origin project. There are others: the late Paul Allen, with his six-engine space plane; Richard Branson, already taking reservations for a Virgin Galactic spacecraft that will carry passengers and satellites into space. It beats trying to build the biggest yacht (though Allen, whose 414-foot *Octopus* has two helipads and a Jet Ski dock, may have held that title, too). Indeed, there's something earnest and boyish about the whole spacefaring effort, something more likeable than busting unions back home on earth. As Bezos put it recently, "If I'm 80 years old and looking back on my life, and I can say that I put in place the heavy-lifting infrastructure that made access to space cheap and inexpensive," then "I'll be a very happy 80-year-old."[1]

Why go to space?

"So that the next generation could have the entrepreneurial explosion like I saw on the internet," said Bezos, conjuring up a vision of brown-and-yellow UPS shuttles delivering printer cartridges to the rings of Saturn.[2] (Sometime this year, Vodafone and Nokia plan to set up a mobile phone network on the moon.)[3]

Or to escape the wreckage of planet Earth. In November 2016, Stephen Hawking told an audience that "spreading out may be the only thing that saves us from ourselves," and gave us a thousand-year timetable to be off the Earth. The following May, he cut the deadline down to a century. "Earth is under threat from so many areas that it is difficult for me to be positive," he said.[4]

Or, most compelling of all, because once you get to space, you're on your own. It's the ultimate libertarian paradise, something lost on none of these visionaries. Consider the physicist Freeman Dyson, who in the late 1950s took a year away from the Institute for Advanced Study in Princeton to help develop a rocket to Saturn that could be powered by a series of nuclear explosions. These plans were abandoned when the Nuclear Test Ban was adopted in 1963, but Dyson remains a space enthusiast because, as he pointed out in 2017, he wants to escape the small patches of the universe "with their laws and treaties and enforcers and tax-gatherers" and wander instead "through the huge stretches of ungovernable wilderness where . . . no bureaucratic authority can be effective."[5] In space, no one can make you pay your taxes.

But little of this is actually going to happen, because that's not how space works. Just as with climate change on earth, physics and biology ultimately rule. Yes, it's possible that we'll be able to mine some rare minerals from passing asteroids, or do some manufacturing in weightlessness, or even establish Musk's colony on Mars. In the scheme of things, however, these are minor accomplishments, unlikely to deflect any of the trends now governing the planet. Everything we learn about life in space makes it clear that we're not going to get a second chance there.

For one, space flight is hard on human beings. Now that a few people have spent a year in orbit, it's become clear the toll it takes on everything from the shape of our eyeballs to the stability of our DNA. As Charles Wohlforth and Amanda Hendrix pointed out shortly after Musk announced his Mars ambitions, just the outward flight would put astronauts at unacceptable risk, as they'd be bombarded with so many cosmic rays from the stars that the risk of cancer would be greater than the risk of a spaceflight accident. Mice exposed to these rays develop "brain damage and cognitive losses" even when they don't get cancer. Here on Earth, water vapor in the atmosphere shields us from these rays, but "it takes two meters of water to filter out about half the radiation, and a cubic meter of water weighs 2,205 pounds. Carrying enough water to insulate a spacecraft is far beyond current capabilities." And it's not just cosmic rays: a 2014 National Academy of Sciences report listed nine health risks for a Mars mission (including heart damage from radiation, food and medicine instability, and poor psychological health) that are at an "unacceptable level."[6]

For another thing, space is endlessly vast. Alpha Centauri, the nearest star to our sun, is 4.37 light years away, which is impossibly far given that the fastest thing we've ever shot into space, a probe called *Helios 2*, which travels one hundred times faster than a bullet, would still take *19,000 years* to get there. Not long before it ran out of fuel and began to hibernate, NASA's planet-hunting *Kepler* satellite tracked down an alien solar system that scientists dubbed Trappist. Its seven Earth-size planets orbit a cooling dwarf star, three of them at distances that might lend themselves to supporting life. It may be the closest possible candidate for a World Like Ours, but it's thirty-nine light years away, which is to say, it would take *Helios* about *180,000 years* to get there, which is to say eighteen times longer than, as we reckon it, human civilization has existed. That's why every science fiction story is filled with worm holes and inexplicable warp drives: they overcome the basic physics of the universe—in books.

At best, we could send transhumans out across the atmosphere. Indeed, some of the AI enthusiasts imagine that's precisely what will happen, arguing that we should be exploring "genetic and/or surgical modifications"[7] to allow for space travel or, more likely, simply sending robots. The Russian tech pioneer Yuri Milner (whose parents named him for Yuri Gagarin, first man in space) is a Silicon Valley mainstay—among other things, he's an investor in the gene-testing company 23andMe (not to mention a partner in Jared Kushner's real estate ventures). In 2017 he announced plans to spend $100 million to send a robot weighing less than a sheet of paper to Alpha Centauri with a giant space sail and a hundred-billion-watt laser. If it works, it will take only twenty years to get the featherweight probe there.

In fact, the very mission I was watching lift off at Cape Canaveral carried the first artificial intelligence into space, an orb called CIMON (Crew Interactive Mobile CompaniON) that had been equipped with the same Watson AI gear that IBM used to win on *Jeopardy!* and beat the world's best Go players. CIMON looks a lot like the original iMac, and in weightlessness, it would float around the space station until summoned, and then use little fans to fly across the capsule and face the astronaut, who could then ask it various technical questions. Before liftoff, a team of chipper Teutonic gents from Airbus, who had developed the orb, talked at some length about how it would offer "partnership and even companionship," and how it would display "infinite patience," and how it would be "like a buddy, like a good friend working together." The engineers who built CIMON had been taking him out to restaurants—when you ask him a question, his dorsal fans help him nod up and down in reply. They invested him with an ISTJ personality on the Myers-Briggs Type scale—that is, he is highly logical. "I don't think anything could be more exciting than launching AI into space for the first time," one of the engineers said. "There's nothing cooler than that," especially given that CIMON has also been trained to eventually spy on its crewmates, examining "group effects that can develop over a long period of time in small teams and that may arise during long-term missions."[8]

Getting people to the moon was incredibly hard, and the moon is 250,000 miles away. But let's say we cross the 50 *million* miles to Mars—then what? To survive, you'd need to go underground. But to what end? *You can go underground on Earth if you want.* And the multibillion-dollar attempts at building a "biosphere" here on our home planet (where building supplies arrived on a truck) ended in abject failure. Kim Stanley Robinson wrote the greatest novels about the colonization of Mars, a trilogy that dates back a quarter century. Now, says their author, he thinks the whole thing would be a mistake. "It creates a moral hazard," he says. People imagine that if we mess up the Earth, we can "always go to Mars or the stars. It's pernicious."[9]

In fact, it's worse than that. It distracts us from the almost unbearable beauty of the planet we already inhabit. In a more recent novel, *Aurora*, Robinson describes a failed mission from Earth to colonize a planet (failed for all the reasons of distance and human frailty I've already described). Some of the colonists actually manage to make it back to Earth, and one, a woman named Freya, born on the spaceship, eventually finds her life's work rebuilding beaches destroyed by the sea-level rise that came with climate change. As the book ends, she's taking her very first swim in the earthly ocean: "Sun beats on her back, the wet strand gleams. Everything is sparking and glary, too bright to look at. A broken wave rushes up the strand, stops, leaves a line of foam." She kneels in the surf, as the outrushing water swirls the sand beneath her legs, "black flecks forming V patterns in tumbling blond grains, sluicing new deltas right before her eyes. What a world. She lets her head down and kisses the sand."[10]

I thought of that sweet ending when I was at Cape Canaveral. The day before the launch, I went on a tour with public affairs officer Greg Harland and SME (subject matter expert) Don Dankert, who had overseen the rebuilding of dunes along the Atlantic shoreline of the Kennedy Space Center. I'd been warned not even to raise the topic of global warming, which was fine with me—I didn't want to get them fired. In any event, there was no need, because the problem was blindingly obvious. We climbed up a small hill overlooking Launch Complex 39, where the Apollo missions left for the moon and where any future Mars mission would likely begin. The ocean was a few hundred yards away—which is perfect in the sense that launching rockets here on the East Coast means that if something goes wrong, they fall into the sea; but not so perfect given that that sea is now rising. NASA started worrying about this sometime after the turn of the century, forming a Dune Vulnerability Team (a DVT, obviously). The worry accelerated dramatically after Hurricane Sandy in 2011. Sandy didn't hit Cape Canaveral—it hit New York City—but even at a distance of a couple of hundred miles, the great storm churned up waves strong enough to break through the barrier of dunes and very nearly swamp the launch complexes. "Dunes that had previously been relatively stable for decades—suddenly they were gone," said University of Florida geologist John Jaeger.

And so those dunes were rebuilt. Dankert had not only found the millions of cubic yards of sand (excavated from a nearby air force base), but he himself planted the last of the 180,000 native shrubs to hold the sand in place. And so far, the new dune has performed, yielding little ground in the face of recent hurricanes. So, perhaps, until a few more chunks of the Antarctic crash into the drink or a bigger storm hits head-on, our escape route to outer space is safe.

But what impressed me more than the new dune was the sheer affection these two men had for the landscape where they worked. "Kennedy Space Center *is* the Merritt Island Wildlife Refuge," said Harland. "We use less than ten percent for our industrial purposes."

"When you look at the beach, it's like 1870s Florida—the longest undisturbed stretch on the Atlantic Coast," Dankert added. "We launch people into space from the middle of a wildlife refuge. That's amazing."

They talked for a long time about their favorite local species: the brown pelicans skimming the ocean just off the beach; the Florida scrub jays; the gopher tortoise. When they were rebuilding the dune, they carefully bucket-trapped and relocated every last one of the tortoises. Before I left, they drove me half an hour across the swamp to a pond near the Space Center headquarters building, just because they wanted to show me some alligators; we could see snouts surfacing near the bank. At each corner of the pond, a sign had been carefully placed: THE ALLIGATORS IN THIS AREA OCCUR HERE NATURALLY. THEY WERE NOT PLACED HERE AND THEY ARE NOT PETS. PUTTING ANY FOOD IN THE WATER FOR ANY REASON WILL CAUSE THEM TO BECOME ACCUSTOMED TO PEOPLE AND POSSIBLY DANGEROUS. The sign continued: if that should happen, it read, THEY MUST BE REMOVED AND DESTROYED.

Something about that sign moved me tremendously. It would have been easy enough to poison the pond, just as it would have been easy enough to bulldoze the gopher tortoise. But NASA didn't, because of a long series of affectionate laws that drew on an emerging understanding of who we are. John Muir, in some ways the first self-conscious Western environmentalist, crossed Florida on his thousand-mile walk from Louisville to the Gulf of Mexico in 1867, a trip he used to form his first heretical thoughts about the meaning of being human. From his diary: "The world, we are told, was made especially for man—a presumption not supported by all the facts. A numerous class of men are painfully astonished whenever they find anything, living or dead, in all God's universe, which they cannot eat or render in some way what they call useful to themselves." His proof that this self-centeredness was wrong was the alligator, numbers of which he could hear roaring in the swamp as he camped nearby, and which clearly caused man mostly trouble. But the alligator was wonderful nonetheless, Muir thought, a remarkable creature perfectly adapted to its landscape. "I have better thoughts of those alligators now that I've seen them at home," he wrote. Indeed, he addressed the creatures directly: "Honorable representatives of the great saurian of an older creation, may you long enjoy your lilies and rushes, and be blessed now and then with a mouthful of terror-stricken man by way of dainty."[11] Most of us don't go as far as Muir—we still wince when we read of some gator emerging from the water hazard on the sixth hole to chomp down on an unwary golfer—but his basic idea that all of creation matters has made some real headway.

That evening, Harland and Dankert drew me a crude map to a beach where I could wait the hours until the predawn rocket launch—a beach where they said I'd be likely to spot a loggerhead sea turtle coming ashore to lay her eggs. And so, I lay on the sand, north of Patrick Air Force Base and south of the sign erected by the Brevard County Historical Commission to commemorate that, here, in 1965, Barbara Eden emerged from her bottle to greet her astronaut at the start of *I Dream of Jeannie* (the last sitcom filmed in black and white, and certainly a key feature of my early intellectual life). The beach was deserted, and under a near-full moon, it was easy to see a turtle trundle from the sea. She lumbered deliberately to a spot near the dune, where she used her powerful legs to excavate a pit. She spent an hour laying eggs, and even from thirty yards away, I could hear her heavy breathing in between the whisper of the waves. And then, having covered her clutch, she tracked back to the ocean, in the fashion of others like her for the last 120 million years.

That humans have made her life harder is undeniable. In some places, sea turtles are eaten; in many more, their habitat has been eaten away, often by beachside cities that in turn foster the raccoons and foxes that delight in digging up turtle eggs. Huge numbers of turtles have been caught up and killed by accident in the hunt for shrimp; in Mexico, three hundred sea turtles were found dead in 2018, trapped in a single abandoned fishing net.[12]

But humans have also now set aside beaches for turtles and have organized patrols to protect their nests—in some places, they cage each nest in wire to keep the raccoons at bay. They've mandated "turtle excluder devices" on shrimp nets. Even the new dune built along the launchpad complex was designed in part to block the lights that often confused the turtles emerging to build their nests. And so, in some places, populations have begun to rebound—only, of course, to be threatened anew by rising heat (the temperature of the sand determines the sex of the eggs) and soaring acidity.

I take two ideas from that turtle nest.

The first is we really do live on an unbearably beautiful planet. We don't think of it often as a planet—we live our daily lives on flat and often prosaic ground, and when we're in the air, the flight attendant usually makes us lower the window shade so as not to interfere with the movie. But even with seven billion of us, the planet remains an astonishing collection not just of cities and suburbs, but of mountains and ice and forests and ocean. I've been to the highest year-round human habitation, the Rongbuk Monastery in Tibet, and stared up from its rocky ground at Everest overhead, its summit so high that it sticks into the jet stream and unrolls a long pennant of white cloud. I've wandered the Antarctic Peninsula, watching glaciers calve icebergs with a thunderous roar. I've climbed on the endless lava fields of Iceland and watched the magma pour into the Pacific Ocean from Kilauea, in Hawaii, birthing new land before my eyes. I've seen the steam puffing from the top of Mount Rainier and wondered if I'd managed to climb it the day it would erupt. And I've lain on my stomach in my backyard, watching beetles wander by, watching dew hang on stalks of grass. I've seen penguins, I've watched whales, I've played with my dog.

We live on a planet—we live on a *planet*. And it's infinitely more glorious than the others we head for at such risk and expense. The single most inhospitable cubic meter of the Earth's surface—some waste of Saharan sand, some rocky Himalayan outcrop—is a thousand times more hospitable than the most appealing corner of Mars or Jupiter. If you wanted for some reason to turn that Saharan desert green, you could do it with some water. You can breathe to the top of the highest peak. Everywhere there is life.

And—this is for me the second lesson—the most curious of all those lives are the human ones, *because we can destroy, but also because we can decide not to destroy.* The turtle does what she does, and magnificently. She can't not do it, though, any more than the beaver can decide to take a break from building dams or the bee from making honey. But if the bird's special gift is flight, ours is the possibility of restraint. We're the only creature who can decide *not* to do something we're capable of doing. That's our superpower, even if we exercise it too rarely.

So, yes, we can wreck the Earth as we've known it, killing vast numbers of ourselves and wiping out entire swaths of other life—in fact, as we've seen, we're doing that right now. But we can also *not* do that. We could instead put a solar panel on the top of every last one of those roofs that I described at the opening of this book, and if we do, then we will have started in a different direction. We can engineer our children, at least a little now and doubtless more in the future—or we can decide *not* to. We can build our replacements in the form of ever-smarter robots, and we can try to keep ourselves alive as digitally preserved consciousnesses—or we can accept with grace that each of us has a moment and a place.

I do not know that we will make these choices. I rather suspect we won't—we are faltering now, and the human game has indeed begun to play itself out. That's what the relentless rise in temperature tells us, and the fact that we increasingly spend our days staring glumly at the rectangle in our palm. But we *could* make those choices. We have the tools (nonviolence chief among them) to allow us to stand up to the powerful and the reckless, and we have the fundamental idea of human solidarity that we could take as our guide.

We are messy creatures, often selfish, prone to short-sightedness, susceptible to greed. In a Trumpian moment with racism and nationalism resurgent, you could argue that our disappearance would be no great loss. And yet, most of us, most of the time, are pretty wonderful: funny, kind. Another name for human solidarity is love, and when I think about our world in its present form, that is what overwhelms me. The human love that works to feed the hungry and clothe the naked, the love that comes together in defense of sea turtles and sea ice and of all else around us that is good. The love that lets each of us see we're not the most important thing on earth, and makes us okay with that. The love that welcomes us, imperfect, into the world and surrounds us when we die.

Even—especially—in its twilight, the human game is graceful and compelling.

NOTES

AN OPENING NOTE ON HOPE

1. Steven Pinker, *Enlightenment Now: The Case for Reason, Science, Humanism, and Progress* (New York: Vintage, 2017), p. 262.

PART ONE: THE SIZE OF THE BOARD

CHAPTER 1

1. Youtu.be/3UgGVKnelfY CertainTeed Roofing, "How Shingles Are Made," youtube.com
2. Yuval Noah Harari, *Homo Deus: A Brief History of Tomorrow* (New York: HarperCollins, 2017), p. 15.
3. Nicholas Kristof, "Good News, Despite What You've Heard," *New York Times*, July 1, 2017.
4. Yuval Noah Harari, *Sapiens: A Brief History of Humankind* (New York: Harper-Collins, 2015), p. 247.
5. Kaushik Basu, "The Global Economy in 2067," *Project Syndicate*, June 21, 2017.
6. "Scientists' Warning to Humanity 'Most Talked about Paper,'" March 7, 2018, sciencedaily.com.
7. Nafeez Ahmed, "NASA-Funded Study: Industrial Civilization Headed for 'Irreversible Collapse'?" *Guardian*, March 14, 2014.
8. Baher Kamal, "Alert: Nature, on the Verge of Bankruptcy," September 12, 2017, ispnews.net.
9. Clive Hamilton, *Defiant Earth: The Fate of Humans in the Anthropocene* (Cambridge, UK: Polity Press, 2017), p. 42.
10. John Vidal, "From Africa's Baobabs to America's Pines: Our Ancient Trees Are Dying," *Huffington Post*, June 19, 2018.
11. Anne Barnard, "Climate Change Is Killing the Cedars of Lebanon," *New York Times*, July 18, 2018.

12. Damian Carrington, "Arctic Stronghold of World's Seeds Flooded After Permafrost Melts," *Guardian*, May 19, 2017.

13. William E. Rees, "Staving Off the Coming Global Collapse," TheTyee.ca, July 17, 2017.

14. Eelco Rohling, *The Oceans: A Deep History* (Princeton, NJ: Princeton University Press, 2017), p. 15.

15. Donella H. Meadows et al., *The Limits to Growth: A Report of the Club of Rome* (New York: Universe Books, 1972), abstract.

16. "A Greener Bush," *The Economist*, February 13, 2003.

17. Peter U. Clark et al., "Consequences of Twenty-First-Century Policy for Multi-Millennial Climate and Sea-Level Change," *Nature Climate Change* 6, no. 4 (February 2016): 360–69.

18. Adam Gopnik, "The Illiberal Imagination," *The New Yorker*, March 20, 2017.

CHAPTER 2

1. Michael Safi, "Pollution Stops Play at Delhi Test Match as Bowlers Struggle to Breathe," *Guardian*, December 3, 2017.

2. Mehreen Zahra-Malik, "In Lahore, Smog Has Become a 'Fifth Season,'" *New York Times*, November 10, 2017.

3. Aniruddha Ghosal, "Landmark Study Lies Buried: How Delhi's Poisonous Air Is Damaging Its Children for Life," *Indian Express*, April 2, 2015.

4. Hilary Brueck, "Pollution Is Killing More People than Wars, Obesity, Smoking, and Malnutrition," *Business Insider*, October 24, 2017.

5. Institute for Governance and Sustainable Development, "Climate Change Could Kill More than 100 Million People by 2030," http://www.igsd.org/climate-change-could-kill-more-than-100-million-people-by-2030/

6. Joe Romm, "Earth's Rate of Global Warming Is 400,000 Hiroshima Bombs a Day," thinkprogress.org, December 22, 2013.

7. Rohling, *The Oceans*, p. 106.

8. Ibid., p. 107.

9. Justin Gillis, "Carbon in Atmosphere Is Rising, Even as Emissions Stabilize," *New York Times*, June 26, 2017.

10. Eric Holthaus, "Antarctic Melt Holds Coastal Cities Hostage. Here's the Way Out," grist.org, June 13, 2018.

11. Harry Cockburn, "Worst Case Climate Change Scenario Could Be More Extreme than Thought, Scientists Warn," *Independent*, May 15, 2018.

12. Damian Carrington, "Record-Breaking Climate Change Pushes World into 'Uncharted Territory,'" *Guardian*, March 20, 2017.

13. Eleanor Cummins, "Tropical Storm Ophelia Really Did Break the Weather Forecast Grid," *Slate*, October 16, 2017.

14. Brett Walton, "Cape Town Rations Water Before Reservoirs Hit Zero," circleofblue.org, October 26, 2017.

15. Samanth Subramanian, "India's Silicon Valley Is Dying of Thirst. Your City May Be Next," *Wired*, May 2, 2017.

16. Marcello Rossi, "In Italy's Parched Po River Valley, Climate Change Threatens the Future of Agriculture," Reuters, July 27, 2017.

17. Catherine Edwards, "The Source of Italy's Longest River Has Dried Up Due to Drought," thelocal.it, September 6, 2017.

18. Chelsea Harvey, "Scientists Find a Surprising Result on Global Wildfires: They're Actually Burning Less Land," *Washington Post*, June 29, 2017.

19. Michael Kodas, *Megafire: The Race to Extinguish a Deadly Epidemic of Flame* (Boston: Houghton Mifflin Harcourt, 2017), p. xii.

20. Ibid., pp. xii, xv.

21. David Karoly, "Bushfires and Extreme Heat in South-East Australia," realclimate.org, February 16, 2009.

22. Regional Municipality of Wood Buffalo, Twitter post, May 4, 2016, 9:28 AM.

23. "Greece Wildfires: Dozens Dead in Attica Region," bbc.com, July 24, 2018.

24. Kodas, *Megafire*, p. 116.

25. Will Dunham, "Bolt from the Blue: Warming Climate May Fuel More Lightning," Reuters, November 13, 2014.

26. Kodas, *Megafire*, p. 20.

27. Jack Healy, "Burying Their Cattle, Ranchers Call Wildfires 'Our Hurricane Katrina,'" *New York Times*, March 20, 2017.

28. Michael E. Mann, "It's a Fact: Climate Change Made Hurricane Harvey More Deadly," *Guardian*, August 28, 2017.

29. Doyle Rice, "Global Warming Makes 'Biblical' Rain Like That from Hurricane Harvey Much More Likely," *USA Today*, November 14, 2017.

30. Ibid.

31. Scott Waldman, "Global Warming Tied to Hurricane Harvey," *Scientific American*, December 14, 2017.

32. Seth Borenstein, "Florence Could Dump Enough to Fill Chesapeake Bay," Associated Press News, September 14, 2018.

33. Somini Sengupta, "The City of My Birth in India Is Becoming a Climate Casualty. It Didn't Have to Be," *New York Times*, July 31, 2018.

34. "Extreme Precipitation Events Have Risen Sharply in Northeastern U.S. Since 1996," *Yale Environment 360*, May 24, 2017.

35. Hiroko Tabuchi et al., "Floods Are Getting Worse, and 2,500 Chemical Sites Lie in the Water's Path," *New York Times*, February 6, 2018.

36. "Glacier Mass Loss: Past the Point of No Return," University of Innsbruck, uibk.ac.at, March 19, 2018.

37. Stephen Leahy, "Hidden Costs of Climate Change Running Hundreds of Billions a Year," *National Geographic*, September 27, 2017.

38. Fiona Harvey, "Climate Change Is Already Damaging World Economy, Report Finds," *Guardian*, September 25, 2012.

39. Richard Harris, "Study Puts Puerto Rico Death Toll from Hurricane Maria Near 5,000," *All Things Considered*, NPR, May 29, 2018.

40. Solomon Hsiang and Trevor Houser, "Don't Let Puerto Rico Fall into an Economic Abyss," *New York Times*, September 29, 2017.

41. Pinker, *Enlightenment Now*, p. 69.

42. "Climate Change Aggravates Global Hunger," Agence France-Presse, September 15, 2017.

43. Lin Taylor, "Factbox: Conflicts and Climate Disasters Forcing Children into Work—U.N.," Reuters.com, June 12, 2018.

44. Laignee Barron, "143 Million People Could Soon Be Displaced Because of Climate Change, World Bank Says," *Time*, March 20, 2018.

45. Daniel Wesangula, "Dying Gods: Mt. Kenya's Disappearing Glaciers Spread Violence Below," *Climate Home News*, August 2, 2017.

46. Lorraine Chow, "The Climate Crisis May Be Taking a Toll on Your Mental Health," *Salon*, May 22, 2017.

47. Ilissa Ocko, "Climate Change Is Messing with Clouds," edf.org/blog, August 24, 2016.

48. Brian Resnick, "We're Witnessing the Fastest Decline in Arctic Sea Ice in at Least 1,500 Years," vox.com, February 16, 2018.

49. Henry Fountain, "Alaska's Permafrost Is Thawing," *New York Times*, August 23, 2017.

50. Gillis, "Carbon in Atmosphere Is Rising."

51. Tom Knudson, "California Is Drilling for Water That Fell to Earth 20,000 Years Ago," *Mother Jones*, March 13, 2015.

52. Carol Rasmussen, "Sierras Lost Water Weight, Grew Taller During Drought," nasa.gov, December 13, 2017.

53. Matt Stevens, "102 Million Dead California Trees 'Unprecedented in Our Modern History,' Officials Say," *Los Angeles Times*, November 18, 2016.

54. Thomas Fuller, "Everything Was Incinerated: Scenes from One Community Wrecked by the Santa Rosa Fire," *New York Times*, October 10, 2017.

55. Andrew Freedman, "The Combustible Mix Behind Southern California's Terrifying Wildfires," mashable.com, December 6, 2017.

56. Nora Gallagher, "Southern Californians Know: Climate Change Is Real, It Is Deadly and It Is Here," *Guardian*, March 3, 2018.

57. Ibid.

CHAPTER 3

1. "Failing Phytoplankton, Failing Oxygen: Global Warming Disaster Could Suffocate Life on Planet Earth," sciencedaily.com, December 1, 2015.

2. Jasmin Fox-Skelly, "There Are Diseases Hidden in Ice and They Are Waking Up," bbc.com, May 4, 2017.

3. Susan Casey, *The Wave: In Pursuit of the Rogues, Freaks, and Giants of the Ocean* (New York: Doubleday, 2010), p. 153.

4. Ibid., p. 253; and Akshat Rathi, "Global Warming Won't Just Change the Weather—It Could Trigger Massive Earthquakes and Volcanoes," qz.com, May 24, 2016.

5. Joe Romm, "Exclusive: Elevated CO_2 Levels Directly Affect Human Cognition, New Harvard Study Shows," thinkprogress.org, October 26, 2015.

6. Anna Vidot, "Climate Change to Blame for Flatlining Wheat Yield Gains: CSIRO," *ABC Rural*, March 8, 2017.

7. Georgina Gustin, "Climate Change Could Lead to Major Crop Failures in World's Biggest Corn Regions," *InsideClimate News*, June 11, 2018.

8. Bill McKibben, "While Colorado Burns, Washington Fiddles," *Guardian*, June 29, 2012.

9. Daisy Dunne, "Global Warming Could Cause Yield of Sorghum Crops to Drop 'Substantially,'" carbonbrief.org, August 14, 2017.

10. Tobias Lunt et al., "Vulnerabilities to Agricultural Production Shocks: An Extreme, Plausible Scenario for Assessment of Risk for the Insurance Sector," *Climate Risk Management* 13 (2016): 1–9.

11. Elizabeth Winkler, "How the Climate Crisis Could Become a Food Crisis Overnight," *Washington Post*, July 27, 2017.

12. Helena Bottemiller Evich, "The Great Nutrient Collapse," *Politico*, September 13, 2017.

13. Brad Plumer, "How More Carbon Dioxide Can Make Food Less Nutritious," *New York Times*, May 23, 2018.

14. Evich, "Great Nutrient Collapse."

15. Bob Berwyn, "Global Warming Means More Insects Threatening Food Crops—a Lot More, Study Warns," *InsideClimate News*, August 30, 2018.

16. http://www.realclimate.org/index.php/archives/2013/10/sea-level-in-the-5th-ipcc-report/

17. Peter Brannen, *The Ends of the World: Volcanic Apocalypses, Lethal Oceans, and Our Quest to Understand Earth's Past Mass Extinctions* (New York: Ecco Books, 2017), p. 258.

18. Robert Scribbler, "New Study Finds That Present CO_2 Levels Are Capable of Melting Large Portions of East and West Antarctica," robertscribbler.com, August 2, 2017.

19. Michael Le Page, "Alarm as Ice Loss from Antarctica Triples in the Past Five Years," *New Scientist*, June 13, 2018.

20. Ian Johnston, "Earth Could Become 'Practically Ungovernable' If Sea Levels Keep Rising, Says Former NASA Climate Chief," *Independent*, July 14, 2017.

21. "How Much Will the Seas Rise?" conversations.e-flux.com, February 26, 2018.

22. David Smiley, "Was Jorge Pérez Drunk When He Made Controversial Sea Level Rise Comment to Jeff Goodell?" *Miami Herald*, May 31, 2018.

23. Jeff Goodell, *The Water Will Come: Rising Seas, Sinking Cities, and the Remaking of the Civilized World* (New York: Little, Brown, and Company, 2017), p. 148.

24. Christopher Flavelle, "Florida Could Be Close to a Real Estate Reckoning," *Insurance Journal*, January 2, 2018.

25. Anna Hirtensen, "AXA Insurance Chief Warns of 'Uninsurable Basements' from New York to Mumbai," *Insurance Journal*, January 26, 2018.

26. Tim Radford, "Kids Suing Trump Get Helping Hand from World's Most Famous Climate Scientist," *EcoWatch*, July 19, 2017.

27. "Relocating Kivalina," toolkit.climate.gov, January 17, 2017.

28. Wallace-Wells, *"The Uninhabitable Earth," New York*, July 9, 2017.

29. Kenneth R. Weiss, "Some of the World's Biggest Lakes Are Drying Up. Here's Why," *Inter Press Service*, March 1, 2018.

30. Bryan Bender, "Chief of U.S. Pacific Forces Calls Climate Greatest Worry," *Boston Globe*, March 9, 2013.

31. Jonathan Watts, "Arctic's Strongest Sea Ice Breaks Up for First Time on Record," *Guardian*, August 21, 2018.

32. Quirin Schiermeier, "Huge Landslide Triggered Rare Greenland Mega-Tsunami," *Nature*, July 27, 2017.

33. Goodell, *The Water Will Come*, p. 141.

CHAPTER 4

1. "How to Improve the Health of the Ocean," *The Economist*, May 27, 2017.

2. Brannen, *Ends of the World*, p. 235.

3. Roz Pidcock, "Rate of Ocean Warming Quadrupled Since Late 20th Century, Study Reveals," *Carbon Brief*, March 10, 2017.

4. Brittany Patterson, "How Much Heat Does the Ocean Trap? A Robot Aims to Find Out," *Climatewire*, October 18, 2016.

5. Christopher Knaus and Nick Evershed, "Great Barrier Reef at Terminal Stage; Scientists Despair at Latest Bleaching Data," *Guardian*, April 9, 2017.

6. Amy Remeikis, "Great Barrier Reef Tourism Spokesman Attacks Scientist Over Slump in Visitors," *Guardian*, January 12, 2018.

7. P. G. Brewer, "A Short History of Ocean Acidification Science in the 20th Century: A Chemist's View," *Biogeosciences* 10 (2013): 7411–22.

8. Rohling, *Oceans*, p. 181.

9. Ibid., p. 72.

10. Ibid., 161.

11. Seth Borenstein, "Scientists Warn of Hot, Sour, Breathless Oceans," Associated Press, November 14, 2013.

12. Elena Becatoros, "More than 90 Percent of World's Coral Reefs Will Die by 2050," *Independent*, March 13, 2017.

13. Brannen, *Ends of the World*, p. 65.

14. Ibid., p. 65.

15. Ibid., p. 122.

16. Ibid., p. 188.

17. Ibid., p. 203.

18. "A One-Two Punch May Have Helped Check the Dinosaurs," sciencedaily.com, February 7, 2018.

19. Joseph F. Byrnes and Leif Karlstrom, "Anomalous K-Pg–aged Seafloor Attributed to Impact-Induced Mid-Ocean Ridge Magmatism," *Science Advances* 4 no. 2 (February 7, 2018): 1–6.

20. Brannen, *Ends of the World*, p. 136.

21. Rohling, *Oceans*, p. 114.

22. Ibid., p. 88.

23. Howard Lee, "Underground Magma Triggered Earth's Worst Mass Extinction with Greenhouse Gases," *Guardian*, August 1, 2017.

24. Damian Carrington, "Earth's Sixth Mass Extinction Event Under Way, Scientists Warn," *Guardian*, July 10, 2017.

25. Damian Carrington, "Humans Just 0.01% of All Life but Have Destroyed 83% of Wild Mammals—Study," *Guardian*, May 21, 2018.

26. George Monbiot, "Our Natural World Is Disappearing Before Our Eyes. We Have to Save It," tppahanshilhorst.com, July 6, 2018.

CHAPTER 5

1. Donald Worster, *Shrinking the Earth: The Rise and Decline of Natural Abundance* (New York: Oxford University Press, 2016), p. 15.

2. Ibid., p. 40.

3. Adam Smith, *The Wealth of Nations*, Book 4, Chapter 7, Part 3. (Indianapolis, 2009), p. 2.

4. Gayathri Vaidyanathan, "Killer Heat Grows Hotter around the World," *Scientific American*, August 6, 2015.

5. Alan Blinder, "As the Northwest Boils, an Aversion to Air-Conditioners Wilts," *New York Times*, August 3, 2017.

6. Mike Ives, "In India, Slight Rise in Temperature Is Tied to Heat Wave Deaths," *New York Times*, June 8, 2017.

7. Jason Samenow, "Two Middle Eastern Locations Hit 129 Degrees, Hottest Ever in Eastern Hemisphere, Maybe the World," *Washington Post*, July 22, 2016.

8. Bob Berwyn, "Heat Waves Creeping Toward a Deadly Heat-Humidity Threshold," *InsideClimate News*, August 3, 2017.

9. Damian Carrington, "Unsurvivable Heatwaves Could Strike the Heart of China by End of Century," *Guardian*, July 31, 2018.

10. Jonathan Watts and Elle Hunt, "Halfway to Boiling: The City at 50C," *Guardian*, August 13, 2018.

11. Kevin Krajick, "Humidity May Prove Breaking Point for Some Areas as Temperatures Rise, Says Study," *Earth Institute*, December 22, 2017.

12. Lauren Morello, "Climate Change Is Cutting Humans' Work Capacity," climatecentral.org, February 24, 2013.

13. Jeremy Deaton, "Extreme Heat Is Killing America's Farm Workers," qz.com, September 1, 2018.

14. Somini Sengupta, Tiffany May, and Zia ur-Rehman, "How Record Heat Wreaked Havoc on Four Continents," *New York Times*, July 30, 2018.

15. Elle Hunt, "'We Have Different Ways of Coping': The Global Heatwave from Beijing to Bukhara," *Guardian*, July 28, 2018.

16. Watts and Hunt, "Halfway to Boiling."

17. Christopher Flavelle, "Louisiana Plan Could Move Thousands from Coast," *Portland Press Herald*, December 22, 2017.

18. Ashley Nagaoka, "Hawaii Study: Impacts of Sea Level Rise Already Being Felt—and It Will Only Get Worse," hawaiinewsnow.com, December 20, 2017.

19. Michael Kimmelman, "Jakarta Is Sinking So Fast, It Could End Up Underwater," *New York Times*, December 21, 2017.

20. Gabrielle Gurley, "Boston's Rendezvous with Climate Destiny," prospect.org, January 5, 2018.

21. Rohling, *Oceans*, p. 170.

22. Goodell, *The Water Will Come*, p. 214.

23. Dr. Jeff Masters, "Retreat from a Rising Sea: A Book Review," wunderground.com/cat6, February 16, 2018.

24. Kavya Balaraman, "U.S. Harvests Could Suffer with Climate Change," *Scientific American*, January 20, 2017.

25. Oliver Millman, "We're Moving to Higher Ground," *Guardian*, September 24, 2018.

26. Worster, *Shrinking the Earth*, p. 133.

27. Alister Doyle, "Arctic Thaw to Cause up to $90 Trillion Damage to Roads and Buildings," *Independent*, April 25, 2017.

28. Catherine Porter, "Canadian Town, Isolated after Losing Rail Link, 'Feels Held Hostage,'" *New York Times*, August 30, 2017.

29. Jim Dwyer, "Saving Scotland's Heritage from the Rising Seas," *New York Times*, September 25, 2018.

CHAPTER 6

1. Nathaniel Rich, "Losing Earth: The Decade We Almost Stopped Climate Change," *New York Times Magazine*, August 1, 2018.

2. Harry Stevens, "A 30-year alarm on the reality of climate change," Axios, June 23, 2018.

3. Peter Frumhoff, "Global Warming Fact: More than Half of All Industrial CO_2 Pollution Has Been Emitted Since 1988," uscusa.org, December 15, 2014.

4. Goodell, *The Water Will Come*, p. 224.

5. Ibid., p. 84.

6. Joe Romm, "Obama's Worst Speech Ever: 'We've Added Enough New Oil and Gas Pipeline to Encircle the Earth,'" thinkprogress.org, March 22, 2012.

7. Sabrina Shankman, "Oil and Gas Fields Leak Far More Methane than EPA Reports, Study Finds," *InsideClimate News*, June 21, 2018.

8. Cameron Cawthorne, "Obama Touts Paris Agreement," *Washington Free Beacon*, November 28, 2018.

9. Nicole Gaouette, "Trudeau Issues Rallying Cry for Climate Fight and Takes a Dig at the US," cnn.com, September 21, 2017.

CHAPTER 7

1. Neela Banerjee, Lisa Song, and David Hasemyer, "Exxon's Own Research Confirmed Fossil Fuels' Role in Global Warming Decades Ago," *InsideClimate News*, September 16, 2015.

2. Ibid.

3. Benjamin Franta, "On Its 100th birthday in 1959, Edward Teller Warned the Oil Industry About Global Warming," *Guardian*, January 1, 2018.

4. Energy and Policy Institute, "Utilities Knew: Documenting Electric Utilities' Early Knowledge and Ongoing Deception on Climate Change from 1968–2017," July 2017.

5. Dick Russell and Robert F. Kennedy Jr., *Horsemen of the Apocalypse: The Men Who Are Destroying Life on Earth—And What It Means for Our Children* (New York: Hot Books, 2017), p. 16–17.

6. Neela Banerjee, Lisa Song, and David Hasemyer, "Exxon: The Road Not Taken," *InsideClimate News*, September 16, 2015.

7. Sara Jerving et al., "What Exxon Knew about the Earth's Melting Arctic," *Los Angeles Times*, October 9, 2015.

8. Benjamin Franta, "Shell and Exxon's Secret 1980s Climate Change Warnings," *Guardian*, September 19, 2018.

9. Jason M. Breslow, "Investigation Finds Exxon Ignored Its Own Early Climate Change Warnings," pbs.org, September 16, 2015.

10. Russell and Kennedy, *Horsemen of the Apocalypse*, p. 20–21.

11. Oliver Burkeman, "Memo Exposes Bush's New Green Strategy," *Guardian*, March 3, 2003.

12. Ruairí Arrieta-Kenna, "Almost 90% of Americans Don't Know There's Scientific Consensus on Global Warming," vox.com, July 6, 2017.

13. Russell and Kennedy, *Horsemen of the Apocalypse*, p. 30.

14. Rupert Neate, "ExxonMobil CEO: Ending Oil Production 'Not Acceptable for Humanity,'" *Guardian*, May 25, 2016.

15. Olivia Beavers, "Trump: Polar Ice Caps Are 'at a Record Level,'" *Hill*, January 28, 2018.

16. John H. Cushman Jr., "Exxon Reports on Climate Risk and Sees Almost None," *InsideClimate News*, February 5, 2018.

17. Exxon Mobile, "Understanding the '#ExxonKnew' controversy," https://corporate.exxonmobil.com/en/key-topics/understanding-the-exxonknew-controversy/understanding-the-exxonknew-controversy/

18. Alex Steffen, "On Climate, Speed Is Everything," *The Nearly Now*, December 7, 2017.

19. Brady Dennis, "Countries Made Only Modest Climate-Change Promises in Paris. They're Falling Short Anyway," *Washington Post*, February 19, 2018.

PART TWO: LEVERAGE

CHAPTER 8

1. David Cole, "Facts and Figures," *New York Review of Books*, July 19, 2018.

2. Philip Alston, "Extreme Poverty in America: Read the UN Special Monitor's Report," *Guardian*, December 15, 2017.

3. Ed Pilkington, "Hookworm, A Disease of Extreme Poverty, Is Thriving in the U.S. South. Why?" *Guardian*, September 5, 2017.

4. Ibid.

5. "Contempt for the Poor in US Drives Cruel Policies, Says UN Expert," ohchr.org, June 4, 2018.

6. Editorial Board, "The Tax Bill that Inequality Created," *New York Times*, December 16, 2017.

7. Noah Kirsch, "The Three Richest Americans Hold More Wealth Than Bottom 50% of the Country, Study Finds," *Forbes*, November 9, 2017.

8. Max Ehrenfreund, "How Trump's Budget Helps the Rich at the Expense of the Poor," *Washington Post*, May 23, 2017.

9. Annie Lowrey, "Jeff Bezos's $150 Billion Fortune Is a Policy Failure," *Atlantic*, August 1, 2018.

10. Les Leopold, *Runaway Inequality: An Activist's Guide to Economic Justice* (New York: Labor Institute Press, 2015), p. 6.

11. Josh Hoxie, "Blacks and Latinos Will Be Broke in a Few Decades," *Fortune*, September 19, 2017.

12. Preeti Varathan, "Millennials Are Set to Be the Most Unequal Generation Yet," qz.com, November 19, 2017.

13. Raj Chetty, interview by Michel Martin, "U.S. Kids Now Less Likely to Earn More than Their Parents," *All Things Considered*, NPR, December 18, 2016.

14. Richard Wilkinson and Kate Pickett, "The Science Is In: Greater Equality Makes Societies Healthier and Richer," evonomics.com, January 26, 2017.

15. Jessica Boddy, "The Forces Driving Middle-Aged White People's Deaths of Despair," *Morning Edition*, March 23, 2017.

16. Tyler Durden, "America's Miserable 21st Century," zerohedge.com, March 4, 2017.

CHAPTER 9

1. "A Very Big Shoe to Fill," *The Economist*, March 7, 2002.

2. Harriet Rubin, "Ayn Rand's Literature of Capitalism," *New York Times*, September 15, 2007.

3. Jonathan Freedland, "The New Age of Ayn Rand: How She Won Over Trump and Silicon Valley," *Guardian*, April 10, 2017.

4. Rubin, "Ayn Rand's Literature of Capitalism."

5. Harriet Rubin, "Fifty Years On, 'Atlas Shrugged' Still Has Its Fans—Especially in Business," *New York Times*, September 17, 2007.

6. Freedland, "New Age of Ayn Rand."

7. Rachel Weiner, "Paul Ryan and Ayn Rand," *Washington Post*, August 13, 2012.

8. Husna Haq, "Paul Ryan Does an About-Face on Ayn Rand," *Christian Science Monitor*, August 14, 2012.

9. Robert James Bidinotto, "Celebrity Ayn Rand Fans," atlassociety.org, January 1, 2006.

10. James B. Stewart, "As a Guru, Ayn Rand May Have Her Limits. Ask Travis Kalanick," *New York Times*, July 13, 2017.

11. Kirsten Powers, "Donald Trump's Kinder, Gentler Version," *USA Today*, April 11, 2016.

12. Wendy Milling, "President Obama Jabs at Ayn Rand, Knocks Himself Out," *Forbes*, October 30, 2012.

13. Jennifer Burns, *Goddess of the Market: Ayn Rand and the American Right* (New York: Oxford University Press, 2009), p. 23.

14. Thomas E. Ricks, *Churchill and Orwell: The Fight for Freedom* (New York: Penguin Press, 2017), p. 8 (emphasis added).

15. William Manchester, *The Last Lion: Winston Spencer Churchill, Alone 1932–1940* (New York: Bantam Books, 1988).

16. George Orwell, *A Patriot After All* (London: Secker and Warburg, 1998), p. 503.

17. Burns, *Goddess of the Market*, p. 8.

18. Ibid., p. 13.

19. Ibid., pp. 20, 24.

20. Anne C. Heller, *Ayn Rand and the World She Made* (New York: Nan A. Talese, 2009), p. 1.

21. Ayn Rand, *The Fountainhead,* twenty-fifth anniversary edition (Indianapolis: Bobbs-Merrill, 1968), p. 7.

22. Ibid., p. 3.

23. Burns, *Goddess of the Market,* p. 86.

24. Ibid.

25. Rand, *The Fountainhead*, p. 712 (emphasis added).

26. Ayn Rand, *Atlas Shrugged* (New York: Dutton, 1957), p. 1065.

27. Andrea Barnet, *Visionary Women: How Rachel Carson, Jane Jacobs, Jane Goodall, and Alice Waters Changed Our World* (New York: Ecco Books, 2018), p. 441.

28. Burns, *Goddess of the Market*, p. 157.

29. Jonas E. Alexis, *Christianity's Dangerous Idea: How the Christian Principle and Spirit Offer the Best Explanation for Life and Why Other Alternatives Fail: Volume 1* (Bloomington, IN: Authorhouse, 2010), p. 600.

CHAPTER 10

1. Maria Tadeo, "Unrepentant Tom Perkins Apologises for 'Kristallnacht' Remarks but Defends War on the Rich Letter," *Independent*, January 28, 2014.

2. Julia Ioffe, "Before Predicting a Liberal Kristallnacht, Tom Perkins Wrote a One-Percenter Romance Novel," *New Republic*, January 25, 2014.

3. Jonathan Chait, "Voting Also Reminds Tom Perkins of Kristallnacht," *New York Magazine*, February 14, 2014.

4. Jane Mayer, "The Koch Brothers Say No to Tariffs," *The New Yorker Radio Hour*, June 15, 2018.

5. Jane Mayer, *Dark Money: The Hidden History of the Billionaires behind the Rise of the Right* (New York: Doubleday, 2016), p. 36.

6. Ibid., p. 38 (emphasis added).

7. Ibid, p. 40.

8. Jane Mayer, "The Secrets of Charles Koch's Political Ascent," *Politico*, January 18, 2016.

9. Nancy MacLean, *Democracy in Chains: The Deep History of the Radical Right's Stealth Plan for America* (New York: Viking, 2017), p. xiv.

10. James M. Buchanan and Gordon Tullock, *The Collected Works of James M. Buchanan, Vol. 3: The Calculus of Consent: Logical Foundations of Constitutional Democracy*, available online at delong.typepad.com/Files/calculus-of-consent.pdf, p. 171.

11. MacLean, *Democracy in Chains*, p. 134.

12. Ibid., p. 148.

13. Mayer, *Dark Money*, p. 464.

14. Jane Mayer, "The Reclusive Hedge-Fund Tycoon Behind the Trump Presidency," *The New Yorker*, March 27, 2017.

15. Lisa Mascaro, "They Snubbed Trump. But the Koch Network Has Still Exerted a Surprising Influence over the White House," *Los Angeles Times*, August 15, 2017.

16. Annie Linskey, "The Koch Brothers (and Their Friends) Want President Trump's Tax Cut. Very Badly," *Boston Globe*, October 14, 2017.

17. Hiroko Tabuchi, "How the Koch Brothers Are Killing Public Transit Projects Around the Country," *New York Times*, June 19, 2018.

18. Lee Gang and Nick Surgey, "Koch Document Reveals Laundry List of Policy Victories Extracted from the Trump Administration," *Intercept*, February 25, 2018.

19. Fredreka Schouten, "Secret Money Funds More than 40% of Outside Congressional Ads," *USA Today*, July 12, 2018.

20. Robert Barnes and Steven Mufson, "White House Counts on Kavanaugh in Battle Against 'Administrative State,'" *Washington Post*, August 12, 2018.

CHAPTER 11

1. Ayn Rand, "Civilization," aynrandlexicon.com.

2. Dhruv Khullar, "How Social Isolation Is Killing Us," *New York Times*, December 22, 2016.

3. Ruth Whippman, "Happiness Is Other People," *New York Times*, October 27, 2017.

4. Nicole Karlis, "Why Doing Good Is Good for the Do-Gooder," *New York Times*, October 26, 2017.

5. "Why Hearing Loss May Raise Your Risk of Dementia," clevelandclinic.org, February 20, 2018.

6. Adam Grant, "In the Company of Givers and Takers," *Harvard Business Review*, April 2013.

7. "Elinor Ostrom," *The Economist*, June 30, 2012.

8. Shankar Vedantam, "Social Isolation Growing in U.S., Study Says," *Washington Post*, June 23, 2006.

9. Jean M. Twenge, *iGen: Why Today's Super-Connected Kids Are Growing Up Less Rebellious, More Tolerant, Less Happy—and Completely Unprepared for Adulthood—and What That Means for the Rest of Us* (New York: Atria Books, 2017).

10. Adam Smith, *The Theory of Moral Sentiments* (London: printed for A. Millar, A. Kincaid, and J. Bell, 1759), part III, chapter 2.

11. Kate Raworth, *Doughnut Economics: Seven Ways to Think Like a 21st-Century Economist* (White River Junction, VT: Chelsea Green Publishing, 2017), p. 89.

12. Mike Bird and Riva Gold, "How Do You Price a Problem Like Korea?" *Wall Street Journal*, August 11, 2017.

13. Michael Tomasky, "The G.O.P.'s Legislative Lemons," *New York Times*, December 14, 2017.

14. Amy Fleming et al., "Heat: The Next Big Inequality Issue," *Guardian*, August 13, 2018.

15. "Rupert Murdoch's Speech on Carbon Neutrality," *Australian*, May 10, 2007.

16. "Free Market Is a Fair Market: Murdoch," *Australian*, April 5, 2013.

17. ClimateDenierRoundup, "Washington Post Hires Former WSJ Opinion Editor, Will He Bring Deniers Along?" dailykos.com, May 31, 2018.

18. Farron Cousins, "Media Matters Report Shows Stunning Lack of Climate Coverage on TV Networks in 2016," desmogblog.com, March 30, 2017.

19. "Fox News' Jesse Watters: 'No One Is Dying from Climate Change,'" Media Matters video, 1:04, mediamatters.org, June 5, 2017.

20. Steven F. Hayward, "Climate Change Has Run Its Course," *Wall Street Journal*, June 4, 2018.

21. Amanda Terkel, "CEI Expert: 'The Best Policy Regarding Global Warming Is to Neglect It,'" thinkprogress.org, August 15, 2006.

22. Robert D. Tollison and Richard E. Wagner, *The Economics of Smoking* (New York: Springer, 1992), p. 183.

23. Graham Readfearn, "The Idea That Climate Scientists Are in It for the Cash Has Deep Ideological Roots," *Guardian*, September 15, 2017.

24. Ibid.

25. Russell and Kennedy, *Horsemen of the Apocalypse*, p. 116.

26. Ibid., p. 114.

27. MacLean, *Democracy in Chains*, p. 216.

28. Bill McKibben, "McCain's Lonely War on Global Warming," *onEarth*, March 31, 2004.

29. Rebecca Shabad, "McCain to Kerry: What Planet Are You On?" *TheHill*, February 19, 2014.

30. Hiroko Tabuchi, "Rooftop Solar Dims Under Pressure from Utility Lobbyists," *New York Times*, July 8, 2017.

31. John Cushman Jr., "No Drop in U.S. Carbon Footprint Expected through 2050, Energy Department Says," *InsideClimate News*, February 6, 2018.

32. Daniel Simmons, "Does the Federal Government Think We Are Dim Bulbs?" instituteforenergyresearch.org, December 18, 2013.

CHAPTER 12

1. Nellie Bowles, "Silicon Valley Flocks to Foiling, Racing Above the Bay's Waves," *New York Times*, August 20, 2017.

2. Adam Vaughan, "Google to Be 100% Powered by Renewable Energy from 2017," *Guardian*, December 6, 2016.

3. Nick Bilton, "Silicon Valley's Most Disturbing Obsession," *Vanity Fair*, November 2016.

4. Maureen Dowd, "Elon Musk's Billion Dollar Crusade to Stop the AI Apocalypse," *Vanity Fair*, April 2017.

5. Melia Robinson, "Silicon Valley's Dream of a Floating, Isolated City Might Actually Happen," *Business Insider*, October 5, 2016.

6. Paulina Borsook, *Cyberselfish: A Critical Romp through the Terribly Libertarian Culture of High Tech* (New York: PublicAffairs, 2000), pp. 2–3.

7. Ibid., p. vi.

8. Ibid., p. 215.

9. Ayn Rand, *Fountainhead*, p. 11.

PART THREE: THE NAME OF THE GAME

CHAPTER 13

1. Personal conversation, November 22, 2017.

2. James Bridle, "Known Unknowns," *Harper's*, July 2018.

3. "Rise of the Machines," *The Economist*, May 22, 2017.

4. "On Welsh Corgis, Computer Vision, and the Power of Deep Learning," microsoft.com, July 14, 2014.

5. Andrew Roberts, "Elon Musk Says to Forget North Korea Because Artificial Intelligence Is the Real Threat to Humanity," uproxx.com, August 12, 2017.

6. Tom Simonite, "What Is Ray Kurzweil Up to at Google? Writing Your Emails," *Wired*, August 2, 2017.

7. Michio Kaku, *The Future of the Mind: The Scientific Quest to Understand, Enhance, and Empower the Mind* (New York: Doubleday, 2014), p. 271.

8. Tim Urban, "What Will Happen When We Succeed in Creating AI That's Smarter than We Are?" qz.com, October 1, 2015.

9. Ibid.

10. Pawel Sysiak, "When Will the First Machine Become Superintelligent?" *Medium*, April 11, 2016.

11. Raffi Khatchadourian, "The Doomsday Invention," *The New Yorker*, November 23, 2015.

12. Tim Urban, "The AI Revolution: The Road to Superintelligence," waitbutwhy .com, January 22, 2015.

CHAPTER 14

1. Kaku, *Future of the Mind*, p. 118.

2. "On Living Forever," interview with Michael West, *Ubiquity* magazine, megafoundation.org, June 2000.

3. Brad Plumer et al., "A Simple Guide to CRISPR, One of the Biggest Science Stories of the Decade," vox.com, July 23, 2018.

4. Ibid.

5. Jennifer A. Doudna and Samuel H. Sternberg, *A Crack in Creation: Gene Editing and the Unthinkable Power to Control Evolution* (Boston: Houghton Mifflin Harcourt, 2017), p. 29.

6. Ibid., p. x.

7. Carl Zimmer, "A Crispr Conundrum: How Cells Fend Off Gene Editing," *New York Times*, June 12, 2018.

8. Doudna and Sternberg, *Crack in Creation*, p. 194.

9. Ibid., p. xv.

10. Ibid., p. 166.

11. Denise Grady, "FDA Panel Recommends Approval for Gene-Altering Leukemia Treatment," *New York Times*, July 12, 2017.

12. Doudna and Sternberg, *Crack in Creation*, p. xvi (emphasis added).

13. Akshat Rathi, "A Highly Successful Attempt at Genetic Editing of Human Embryos Has Opened the Door to Eradicating Inherited Diseases," qz.com, August 2, 2017.

14. Dennis Normille, "CRISPR Bombshell: Chinese Researcher Claims to Have Created Gene-Edited Twins," *Science*, November 26, 2018.

15. "Chinese Scientist Pauses Gene-Edited Baby Trial After Outcry," *Al-Jazeera*, November 28, 2018.

16. Hannah Devlin, "Jennifer Doudna: I Have to be True to Who I Am as a Scientist," *Guardian*, July 2, 2017.

17. Paul Knoepfler, *GMO Sapiens: The Life-Changing Science of Designer Babies* (Singapore: World Scientific Publishing, 2016), p. 11.
18. Dean Hamer, "Tweaking the Genetics of Behavior," *Scientific American*, Fall 1999, p. 62.
19. Gregory E. Pence, *Who's Afraid of Human Cloning?* (Lanham, MD: Rowman and Littlefield, 1998), p. 168.
20. Abbey Interrante, "A New Genetic Test Could Help Determine Children's Success," *Newsweek*, July 10, 2018.
21. Pam Belluck, "Gene Editing for 'Designer Babies'? Highly Unlikely, Scientists Say," *New York Times*, August 4, 2017.
22. Ibid.
23. Ibid.
24. Doudna and Sternberg, *Crack in Creation*, p. 241.
25. Ibid., p. 185.
26. Lee M. Silver, *Remaking Eden* (New York: 1997), pp. 1–3.
27. Doudna and Sternberg, *Crack in Creation*, p. xvi.

CHAPTER 15

1. Abate, T. "Nobel Winner's Theories Raise Uproar in Berkeley/Geneticist's Views Strike Many as Racist, Sexist," *San Francisco Chronicle*, November 13, 2000. Retrieved on October 24, 2007.
2. Doudna and Sternberg, *Crack in Creation*, p. 199.
3. Ibid., p. 237.
4. Silver, *Remaking Eden*, p. 241.
5. Julian Savulescu, "As a Species, We Have a Moral Obligation to Enhance Ourselves," interview by TED guest author, ideas.ted.com, February 19, 2014.
6. Nathaniel Comfort, "Can We Cure Genetic Diseases without Slipping into Eugenics?" *The Nation*, July 16, 2015.
7. Johann Hari, "Is Neoliberalism Making Our Depression and Anxiety Crisis Worse?" *In These Times*, February 21, 2018.
8. Vince Beiser, "The Robot Assault on Fukushima," *Wired*, April 26, 2018.
9. Quoctrung Bui, "Bricklayers Fending Off a Robot Takeover," *New York Times*, March 9, 2018.
10. Harari, *Homo Deus*, p. 330.
11. Alana Semuels, "Where Automation Poses the Greatest Threat to American Jobs," citylab.com, May 3, 2017.
12. Tom Price, "The Last Auto Mechanic," *Medium*, July 27, 2017.
13. Tyler Cower, *Average Is Over* (New York: Dutton, 2013) p. 23.
14. Curtis White, *We, Robots: Staying Human in the Age of Big Data* (Brooklyn, NY: Melville House, 2015), p. 19.
15. Kai-Fu Lee, "The Real Threat of Artificial Intelligence," *New York Times*, June 24, 2017.
16. Bill Joy, "Why the Future Doesn't Need Us," *Wired*, April 1, 2000.
17. Sarah Marsh, "Essays Reveal Stephen Hawking Predicted Race of Superhumans," *Guardian,* October 4, 2018
18. Dowd, "Elon Musk's Billion Dollar Crusade."

19. James Vincent, "Elon Musk Says We Need to Regulate AI Before It Becomes a Danger to Humanity," theverge.com, July 17, 2017.

20. Stephen Hawking, "Artificial Intelligence Could Be the Greatest Disaster in Human History," *Independent*, October 20, 2016.

21. James Barrat, *Our Final Invention: Artificial Intelligence and the End of the Human Era* (New York: St. Martin's Press, 2013), p. 34.

22. Nick Bostrom, "A Transhumanist Perspective on Genetic Enhancements," nickbostrom.com, 2003.

23. Khatchadourian, "Doomsday Invention."

24. Stephen M. Omohundro, "The Basic A.I. Drives," in *Artificial General Intelligence 2008*, eds. Pei Wang, Ben Goertzel, and Stan Franklin (Amsterdam: IOS Press, 2008), available online at selfawaresystems.files.wordpress.com/2008/01/ai_drives_final.pdf, p. 9.

25. Anders Sandberg, "Why We Should Fear the Paperclipper," sentientdevelopments.com, February 14, 2011.

26. Dowd, "Elon Musk's Billion Dollar Crusade," p. 89.

27. Barrat, *Our Final Invention*, p. 19.

28. Ibid., p. 265.

29. Pinker, *Enlightenment Now*, p. 300.

30. Jaron Lanier, *Ten Arguments for Deleting Your Social Media Accounts Right Now* (New York: Henry Holt, 2018), p. 135.

31. Damien Cave, "Artificial Stupidity," *Salon*, October 4, 2000.

32. Dowd, "Elon Musk's Billion Dollar Crusade," p. 90.

33. Sam Thielman, "Is Facebook Even Capable of Stopping an Influence Campaign on Its Platform?" *Talking Points Memo*, September 15, 2017.

34. James Walker, "Researchers Shut Down AI that Invented Its Own Language," digitaljournal.com, July 21, 2017.

35. Cade Metz, "Mark Zuckerberg, Elon Musk, and the Feud over Killer Robots," *New York Times*, June 9, 2018.

36. Dowd, "Elon Musk's Billion Dollar Crusade," p. 91.

37. Khatchadourian, "Doomsday Invention."

CHAPTER 16

1. Knoepfler, *GMO Sapiens*, p. 177.

2. Rachel Nuwer, "Babies Start Learning Language in the Womb," smithsonianmag.com, January 4, 2013.

3. Michael D. Lemonick, "Designer Babies," *Time*, January 11, 1999.

4. Jim Kozubek, "Can Crispr-Cas9 Boost Intelligence?" *Scientific American*, September 23, 2016.

5. Knoepfler, *GMO Sapiens*, p. 179.

6. Ibid., p. 187.

7. Gregory Stock, *Redesigning Humans* (New York, 2002), p. 120.

8. Ephrat Livni, "Columbia and Yale Scientists Found the Spiritual Part of Our Brain," qz.com, May 30, 2018.

9. Knoepfler, *GMO Sapiens*, p. 214.

10. Ray Kurzweil, "Kurzweil's Law," longnow.org, September 23, 2005.

11. Megan Molteni, "Extra CRISPR," *Wired*, May 2018

12. Mihalyi Csikszentmihalyi, *Beyond Boredom and Anxiety, 25th Anniversary Edition* (San Francisco, CA: Wiley and Co., 2000), p. 33.

CHAPTER 17

1. Gil Press, "Breaking News: Humans Will Forever Triumph over Machines," *Forbes*, June 30, 2015.

2. Timothy J. Demy and Gary P. Stewart, eds., *Genetic Engineering: A Christian Response: Crucial Considerations for Shaping Life* (Grand Rapids, MI: Kregel, 1999), p. 131.

3. Wesley J. Smith, "Darwinist Wants Us to Create 'Humanzee,'" nationalreview .com, March 8, 2018.

4. Ed Regis, *The Great Mambo Chicken and the Transhuman Condition* (New York: Basic, 1990), p. 167.

5. Tim Urban, "The AI Revolution: The Road to Superintelligence," *Huffington Post*, February 10, 2015.

6. Ibid.

7. Decca Aitkenhead, "James Lovelock: Before the End of This Century, Robots Will Have Taken Over," *Guardian*, September 30, 2016.

8. Yuval Harari, "The Meaning of Life in a World without Work," *Guardian*, May 8, 2017.

9. Samuel Gibbs, "Apple Co-founder Steve Wozniak Says Humans Will Be Robots' Pets," *Guardian*, June 25, 2015.

10. Paul Lewis, "'Our Minds Can Be Hijacked': The Tech Insiders Who Fear a Smartphone Dystopia," *Guardian*, October 6, 2017.

11. Lanier, *Ten Arguments*, p. 18.

12. Sang In Jung et al., "The Effect of Smartphone Usage Time on Posture and Respiratory Function," *Journal of Physical Therapy Science* 28, no. 1 (January 2016).

13. "The Next Human: Taking Evolution into Our Own Hands," *National Geographic*, April 2017.

14. Jean M. Twenge, "Have Smartphones Destroyed a Generation?" *The Atlantic*, September 2017.

CHAPTER 18

1. Steven Johnson, *How We Got to Now: Six Innovations that Made the Modern World* (New York: Riverhead Books, 2014), p. 148.

2. Jason Pontin, "Silicon Valley's Immortalists Will Help Us All Stay Healthy," *Wired*, December 15, 2017.

3. Maya Kosoff, "Peter Thiel Wants to Inject Himself with Young People's Blood," *Vanity Fair*, August 1, 2016.

4. Maya Kosoff, "This Anti-Aging Startup Is Charging Thousands of Dollars for Teen Blood," *Vanity Fair*, June 1, 2017.

5. Peter Thiel, "The Education of a Libertarian," cato-unbound.org, April 13, 2009.

6. Benjamin Snyder, "This Google Exec Says We Can Live to 500," *Fortune*, March 9, 2015.

7. Katrina Brooker, "Google Ventures and Bill Maris' Search for Immortality," stuff.co.nz, March 11, 2015.

8. Sy Mukherjee, "We're Finally Learning More Details about Alphabet's Secretive Anti-Aging Startup Calico," *Fortune*, December 14, 2017.

9. Nikhil Swaminathan, "A Silicon Valley Scientist and Entrepreneur Who Invented a Drug to Explode Double Chins Is Now Working on a Cure for Aging," qz .com, January 6, 2017.

10. Jamie Nimmo, "Life . . . UNLIMITED: Beating Ageing Is Set to Become the Biggest Business in the World, Say Tycoons," thisismoney.co.uk, March 17, 2018.

11. Zack Guzman, "This Company Will Freeze Your Dead Body for $200,000," nbcnews.com, April 26, 2016.

12. Mark O'Connell, *To Be a Machine* (New York: Doubleday, 2017), p. 23.

13. Antonio Regalado, "A Startup Is Pitching a Mind-Uploading Service that Is '100 Percent Fatal,'" *MIT Technology Review*, March 13, 2018.

14. Ray Kurzweil, *The Age of Spiritual Machines: When Computers Exceed Human Intelligence* (New York: Viking, 1999), p. 97.

15. Tad Friend, "Silicon Valley's Quest to Live Forever," *The New Yorker*, April 3, 2017.

16. Sage Crossroads, *The Fight over the Future: A Collection of Sage Crossroads Debates that Examine the Implications of Aging-Related Research* (Bloomington, IN: iUniverse, 2004), p. 25.

PART FOUR: AN OUTSIDE CHANCE

CHAPTER 19

1. Address of His Holiness Pope Francis, June 7, 2018 wz.vatican.va/Francesco/en /speeches/2018/june/documents/popa.francesco_20180609_imprenditori-energia.html

2. Maggie Astor, "Want to Be Happy? Try Moving to Finland," *New York Times*, March 14, 2018.

3. "Few Americans Support Cuts to Most Government Programs, Including Medicaid," Pew Research Center, Washington, DC, May 26, 2017.

4. Doudna and Sternberg, *Crack in Creation*, p. 234.

5. Lester Thurow, *Creating Wealth: Building the Wealth Pyramid for Individuals, Corporations, and Society* (New York: Nicholas Brealey Publishing, 1999), p. 33.

6. Emily Baumgartner, "As D.I.Y. Gene Editing Gains Popularity, 'Someone Is Going to Get Hurt,'" *New York Times*, May 14, 2018.

7. Maxwell Mehlman, "Regulating Genetic Enhancement," *Wake Forest Law Review* 34 (Fall 1999): 714.

8. Eugene Volokh, "If It Becomes Possible to Safely Genetically Increase Babies' IQ, It Will Become Inevitable," *Washington Post*, July 14, 2015.

9. Daniela Hernandez, "How to Survive a Robot Apocalypse: Just Close the Door," *Wall Street Journal*, November 10, 2017.

10. Olivia Solon, "The Rise of Pseudo-AI: How Tech Firms Quietly Use Humans to Do Bots' Work," *Guardian*, July 6, 2018.

11. James Vincent, "Elon Musk Says We Need to Regulate AI Before It Becomes a Danger to Humanity," theverge.com, July 17, 2017.

12. Preetika Rana, "China, Unhampered by Rules, Races Ahead in Gene-Editing Trials," *Wall Street Journal*, January 21, 2018.

13. "Biggest AI Startup Boosts Fundraising to $1.2 Billion," *Bloomberg News*, May 30, 2018.

14. James Vincent, "Putin Says the Nation that Leads in AI 'Will Be the Ruler of the World,'" theverge.com, September 4, 2017.

15. Maureen Dowd, "Will Mark Zuckerberg 'Like' This Column?" *New York Times*, November 23, 2017.

16. Susan Ratcliffe, ed., "J. Robert Oppenheimer 1904–67, American Physicist," oxfordreference.com, 2016.

17. Barrat, *Our Final Invention*, p. 52.

18. Julian Savulescu, "As a Species, We Have a Moral Obligation to Enhance Ourselves," ideas.ted.com, February 19, 2014.

19. Ingmar Persson and Julian Savulescu, *Unfit for the Future: The Need for Moral Enhancement* (Oxford: Oxford University Press, 2012), pp. 1–2.

20. Ibid., p. 116.

21. David Roberts, "Americans Are Willing to Pay $177 a Year to Avoid Climate Change," vox.com, October 13, 2017.

CHAPTER 20

1. Worster, *Shrinking the Earth*, p. 116.

2. World Bank, "State of Electricity Access Report (SEAR) 2017," worldbank.org.

3. Russell and Kennedy, *Horsemen of the Apocalypse*, pp. 109–10.

4. Ryan Koronowski, "Exxon CEO: What Good Is It to Save the Planet If Humanity Suffers?" thinkprogress.org, May 30, 2013.

5. Simon Evans, "Renewables Will Give More People Access to Electricity than Coal, Says IEA," carbonbrief.org, October 19, 2017.

6. David Roberts, "Wind Power Costs Could Drop 50%. Solar PV Could Provide up to 50% of Global Power. Damn," vox.com, August 31, 2017.

7. Jake Richardson, "Solar Power Energy Payback Time Is Now Super Short," cleantechnica.com, March 25, 2018.

8. Lorraine Chow, "100% Renewable Energy Worldwide Isn't Just Possible—It's Also More Cost-Effective," ecowatch.com, December 22, 2017.

9. Personal conversation with author, September 22, 2016.

10. Natasha Geiling, "New Study Gives 150 Million Reasons to Reduce Carbon Emissions," March 20, 2018, thinkprogress.org.

11. Steve Hanley, "Network of Tesla Powerwall Batteries Saves Green Mountain Power $500,000 During Heat Wave," cleantechnica.com, July 27, 2018.

12. Nick Harmsen, "Elon Musk's Tesla and SA Labor Reach Deal to Give Solar Panels and Batteries to 50,000 Homes," abc.net.au, February 3, 2018.

CHAPTER 21

1. Henry David Thoreau, *On the Duty of Civil Disobedience.* Constitution.org/civ/civildis.htm

2. Jonathan Schell, *The Unconquerable World: Power, Nonviolence, and the Will of the People* (New York: Metropolitan Books, 2003), p. 144. (Emphasis mine.)

3. Harari, *Homo Deus*, p. 277.

4. Heidi M. Przybyla, "Report: Anti-Protester Bills Gain Traction in State Legislatures," *USA Today*, August 29, 2017.

5. Megan Darby, "Pope Francis Tells Oil Chiefs to Keep It in the Ground," climatechangenews.com, June 9, 2018.

6. Phil McKenna, "Ranchers Fight Keystone XL Pipeline by Building Solar Panels in Its Path," *InsideClimate News*, July 11, 2017.

CHAPTER 22

1. "Poll: Voters in America's Heartland Don't Want Changes to National Monuments," nationalparkstraveler.org, November 7, 2017.

2. Coral Davenport and Lisa Friedman, "GOP Pushes to Overhaul Law Meant to Protect At-Risk Species," *New York Times*, July 22, 2018.

3. Wendell Berry, "Manifesto: The Mad Farmer Liberation Front," *In Context* 30 (Fall/Winter 1991).

4. "75 Percent of the German Energy Coops Finance with Local Coop Banks," Die Genossenschaften, dgrv.de https://www.dgrv.de/en/services/energycooperatives /energycoopsfinancewithlocalcoopbanks.html

5. Christine Emba, "Our Socialist Youth: Why Millennials Are Embracing a Bad, Old Term," *Washington Post*, March 21, 2016.

6. David Fleming, ed. Shaun Chamberlin, *Surviving the Future: Culture, Carnival, and Capital in the Aftermath of the Market Economy* (White River Junction, VT: Chelsea Green Publishing, 2016), p. 27.

7. Clint Carter, "We Will Not Get Bigger, We Will Not Get Faster," medium.com, July 26, 2018.

8. Adrien Marck et al., "Are We Reaching the Limits of Homo Sapiens," *Frontiers in Physiology*, October 24, 2017.

9. Richard Price, "Stephen Pinker's Enlightenment Now: the Flynn Effect," richardprice.io, April 6, 2018.

10. Rory Smith, "IQ Scores Are Falling and Have Been for Decades, New Study Finds," CNN.com, June 14, 2018.

11. Adrien Marck et al., "Are We Reaching the Limits of *Homo sapiens*?" *Frontiers in Physiology*, frontiersin.org, October 24, 2017.

12. Steven Pinker, "The Moral Imperative for Bioethics," *Boston Globe*, July 31, 2015.

13. Derrick O'Keefe, "Décroissance in America: Say Degrowth!" *Reporterre*, May 8, 2010.

14. Adam Smith, *The Wealth of Nations*, Book I, chapter 9. "On the Profits of Stock," available online econolib.org/library/smith/smwn.htm

15. John Stuart Mill, "Of the Stationary State of Wealth and Population," quoted at bartleby.com.

16. The Arts Council of Great Britain, "First Annual Report 1945–6" (London: Baynard Press, 1946), p. i.

CHAPTER 23

1. Clay Routledge, "Suicides Have Increased. Is This an Existential Crisis?" *New York Times*, June 23, 2018.

2. Ibid.

3. Edward Luce, *The Retreat of Western Liberalism* (New York: Atlantic Monthly Press, 2017), p. 123.

4. Steven Overly, "Trump's Pick for Labor Secretary Has Said Machines Are Cheaper, Easier to Manage than Humans," *Washington Post*, December 8, 2016.

EPILOGUE: GROUNDED

1. Christian Davenport, "Jeff Bezos on Nuclear Reactors in Space, the Lack of Bacon on Mars and Humanity's Destiny in the Solar System," *Washington Post*, September 15, 2016.

2. Ibid.

3. Arjun Kharpal, "When Elon Musk Sends People to the Moon There May Be a Mobile Network So They Can Check Facebook," cnbc.com, March 2, 2018.

4. Ben Guarino, "Stephen Hawking Calls for a Return to Moon as Earth's Clock Runs Out," *Washington Post*, June 21, 2017.

5. Freeman Dyson, "Should Humans Colonize Space?" Letters, *New York Review of Books*, May 25, 2017.

6. Charles Wohlforth and Amanda Hendrix, "Humans May Dream of Traveling to Mars, but Our Bodies Aren't Built for It," *Los Angeles Times*, November 28, 2016.

7. Rae Paoletta, "Will Human Beings Have to Upgrade Their Bodies to Survive on Mars?" gizmodo.com, March 17, 2017.

8. Sarah Scoles, "The Floating Robot with an IBM Brain Is Headed to Space," *Wired*, June 28, 2018.

9. "The 12 Greatest Challenges for Space Exploration," *Wired*, February 16, 2016.

10. Kim Stanley Robinson, *Aurora* (New York: Orbit, 2015), p. 501.

11. John Muir, *The Ten Thousand Mile Walk to the Gulf,* chapter 5, available online vault.sierraclub.org

12. Luke Bailey, "Three Hundred Turtles Were Found Dead in an Old Fishing Net," inews.co.uk, September 4, 2018.

ACKNOWLEDGMENTS

This book is dedicated to Koreti Tiumalu, a gifted activist and a dear friend, who died much too early, in 2017. I hope that as her young son, Viliamu, grows, he will know how much his mother meant to the climate fight. The dedication is broader than that, however—it extends to all the climate activists I've had the chance to work with over the years. I can't begin to say how much their willingness to fight means to me: every one of them realizes that this is a battle against long odds, with no guarantee of victory (indeed, a guarantee of at least some defeats), and yet they persevere with creativity and passion and love. I spend the most time with my colleagues at 350.org, of course, and it has been a great privilege to watch the young people who launched it grow into full adulthood—the weddings, and the birth announcements, are happy days on the calendar every year. To work with people whom one loves and admires is a great privilege. Day to day, my ace colleague Vanessa Arcara keeps me going.

Naomi Klein, Jane Mayer, and Rebecca Solnit read this book in draft form; each of them has been instrumental in shaping my thinking over the years, and the world owes them great thanks for their reporting and writing. Marcy Darnovsky also brought her sharp eye to bear on the chapters about genetic engineering, for which I am very grateful. Marcy and her colleague Rich Hayes helped me extensively fifteen years ago when I wrote a book on human genetic engineering called *Enough*; I've crossed my own tracks several times in this volume. And I've relied on the work of many others throughout: *New Yorker* reporters Tad Friend and Raffi Khatchadourian have provided superb coverage of Silicon Valley; Dick Russell has helped chronicle the saga of the oil companies and their fight to ward off climate action; Anne Heller and Jennifer Burns produced invaluable biographies of Ayn Rand. There is now a superb community of climate reporters, on the web and in print, which makes me feel very much less lonely than I did thirty years ago. In particular, day-in day-out climate reporting of the *Guardian* and the *New York Times* has been exceptional for years now; the eon-in eon-out geological reporting of Peter Brannen, Eelco Rohling, and Elizabeth Kolbert has broadened my understanding of this particular moment in time.

The New Yorker excerpted this book—thanks to Emily Stokes for her superb editing—and it provided support for much of the reporting that ended up in these pages, from Arizona to Africa to Australia. Thanks to all its fact-checkers and copy editors, and thanks most of all to David Remnick.

I am very grateful to my colleagues at Middlebury College, especially Laurie Patton, Nan Jenks-Jay, Janet Wiseman, Mike Hussey, and Jon Isham, and to the fine students who help keep me on my toes. Our neighbors in Vermont—Warren and Barry King above all—are crucial parts of my life.

I've had the good luck to have the same publishers for many years now—Paul Golob, Maggie Richards, Marian Brown, Caroline Wray, Fiora Elbers-Tibbitts, Austin Price, and their colleagues at Henry Holt always make my books much better. Since in some ways this book stems from *The End of Nature*, I'd like to thank again the two people who made that book a success: David Rosenthal and Annik Lafarge. And I've had the same agent all those decades: Gloria Loomis is a bulwark and friend, and now her assistant, Julia Masnik, plays a major role as well.

When people ask why I keep fighting, one clear answer is my daughter, Sophie. Her mother, Sue Halpern, is my great friend on earth, and constant companion.

INDEX

ABOUT THE AUTHOR

Bill McKibben is a founder of the environmental organization 350.org and was among the early advocates for action on global warming. He is the author of seventeen books, including the best sellers *The End of Nature*, *Eaarth*, and *Deep Economy.* He is the Schumann Distinguished Scholar in Environmental Studies at Middlebury College and the winner of the Gandhi Peace Award, the Thomas Merton Award, and the Right Livelihood Prize. He lives in Vermont.